OXFORD
UNIVERSITY PRESS

C000043632

ESSENTIAL

MATHEMATICS STAGE 8

FOR CAMBRIDGE SECONDARY 1

WORKBOOK

Andrew Manning

Patrick Kivlin, Sue Pemberton, Paul Winters

Great Clarendon Street, Oxford, OX2 6DP, United Kingdom

Oxford University Press is a department of the University of Oxford.
It furthers the University's objective of excellence in research, scholarship,
and education by publishing worldwide. Oxford is a registered trade mark of
Oxford University Press in the UK and in certain other countries

First published in 2014

British Library Cataloguing in Publication Data
Data available

978-1-4085-1987-5

10 9 8

Printed and bound by CPI Group (UK) Ltd, Croydon, CR0 4YY

Acknowledgements

Page make-up: OKS Prepress, India
Illustrations: OKS Prepress, India

The publishers would like to thank the following for permissions to
use their photographs:

Cover: aricvyhmeister/iStockphoto

Although we have made every effort to trace and contact all
copyright holders before publication this has not been possible in all
cases. If notified, the publisher will rectify any errors or omissions at
the earliest opportunity.

Links to third party websites are provided by Oxford in good faith
and for information only. Oxford disclaims any responsibility for
the materials contained in any third party website referenced in
this work.

Contents

Introduction

Welcome to *Mathematics for Cambridge Secondary 1!* This workbook has been written for the Cambridge International Examinations Secondary 1 Mathematics Curriculum Framework and provides complete coverage of Stage 8. Created specifically for international students and teachers by a dedicated and experienced author team, this book covers all areas in the curriculum: number, algebra, geometry, handling data and problem solving.

Using this workbook

This 'write-in' workbook contains a comprehensive range of engaging and stimulating exercises to help you to develop the skills and understanding you need to succeed in Maths. It is divided into exercises that support the topics in the Stage 8 student book, and is arranged in the same order. You will find that looking back at the student book helps you to complete the exercises.

Your teacher can set these exercises as homework or in-class activities. Your teacher may ask you to work alone, in pairs or in groups. For some exercises, your teacher may ask you to report back to your class.

There is a clear description of what the exercise will help you to learn at the start. There is space for you to write your answers, and for your teacher to mark your work and give you some advice and feedback. Your workbook can become a helpful record of your progress.

Some questions are marked with the following symbols:

 Problem solving questions: Help develop knowledge and skills by requiring creative or methodical approaches, often in a real-life context.

 Extension questions: Provide you with further challenge beyond the standard questions found in the book.

Extension chapters: Chapters 21 and 22 are included in the Cambridge IGCSE curriculum. This additional content is included to challenge more able students.

Your teacher will be able to find answers to the exercises on their teacher's CD-ROM.

Chapter 1 Integers, powers and roots

Working with integers

Student book topic 1.1

$$-2 + (+5) = -2 + 5 = 3$$
$$-2 + (-5) = -2 - 5 = -7$$
$$-2 - (+5) = -2 - 5 = -7$$
$$-2 - (-5) = -2 + 5 = 3$$

$$(+3) \times (+4) = 12$$
$$(+3) \times (-4) = -12$$
$$(-3) \times (+4) = -12$$
$$(-3) \times (-4) = 12$$

1 Complete these walls by adding the numbers in two bricks to get the number in the brick above them.

a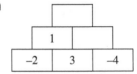

b

c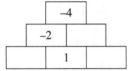

2 Fill in the missing numbers to complete these questions and answers.

a $(-6) + \underline{\quad} = (-3)$ **b** $\underline{\quad} \times (-3) = (-12)$ **c** $(-4) \times \underline{\quad} = 20$

d $(-5) - \underline{\quad} = (-12)$ **e** $\underline{\quad} - (-3) = 1$ **f** $\underline{\quad} \times (-3) = 3$

3 Complete these diagrams. On each line, the numbers in the circles multiply to the numbers in the squares.

a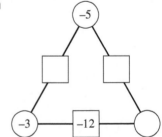

b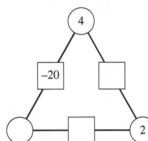

c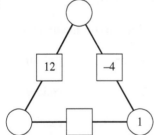

4 Draw arrows from each question on the top to a question on the bottom with the same answer.

| $(-4) + 8$ | $8 + (-6)$ | $(-1) - (-6)$ | $(-7) - (-3)$ | $7 + (-2)$ | $(-4) \times 0$ |

| $(-5) \times (-1)$ | $(-2) - (-2)$ | $(-2) \times (-2)$ | $(-1) \times 4$ | $(-2) - (-4)$ | $(-2) + 7$ |

Exercise 2
Division and mental strategies
Student book topic 1.2

positive ÷ positive = positive

positive ÷ negative = negative

negative ÷ positive = negative

negative ÷ negative = positive

1 Complete these questions and answers.

a $(-12) \div 3 =$ _____

b $15 \div (-5) =$ _____

c $(-12) \div$ _____ $= -2$

d _____ $\div 6 = (-3)$

e $(-20) \div$ _____ $= 4$

f _____ $\div -1 = -5$

2 Match the questions on the top line with the answers on the bottom line.

| $(-3) \div (-1)$ | $(-15) \div 3$ | $0 \div (-5)$ | $6 \div (-2)$ | $(-9) \div (-1)$ | $30 \div 6$ | $36 \div (-4)$ |

| 0 | 5 | -5 | -9 | 3 | 9 | -3 |

3 Use the fact that $456 \div 24 = 19$ to complete these:

a $(-456) \div 24 =$ _____

b $4560 \div (-24) =$ _____

c $456 \div$ _____ $= (-24)$

d $(-456) \div$ _____ $= 19$

e $(-456) \div (-12) =$ _____

f $(-19) \times$ _____ $= 456$

4 Colour in the clown using this key:

Answer	Colour
-5	yellow
-4	black
-3	red
3	green
4	blue

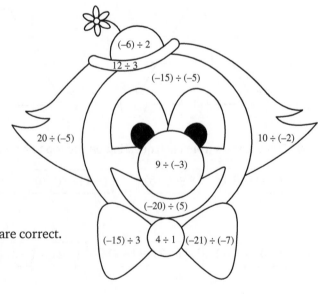

5 Complete the grid below so that all six answers are correct.

-72 ÷ -6 = _____

÷ ÷ ÷

_____ ÷ _____ = _____

= = =

-12 ÷ _____ = -6

The order of operations (BIDMAS):

Brackets

Indices or powers

Division and
Multiplication

Addition and
Subtraction

1 Complete these.

a $(6 + 2) \times 4 =$ _____

b $24 \div (8 - 2) =$ _____

c $5 + (3 - 1) =$ _____

d $(-4 + 3) \times 2 =$ _____

e $-15 + (3 - 12) =$ _____

f $-32 \div (1 - 5) =$ _____

2 Some of these pairs of calculations have the same answer.

Some have different answers.

Put a ✓ next to the pairs with the same answer and a ✗ next to the pairs with different answers.

a $(12 + 6) + 3$ and $12 + (6 + 3)$

b $(15 - 7) - 2$ and $15 - (7 - 2)$

c $(6 \times 2) \times 4$ and $6 \times (2 \times 4)$

d $(24 \div 4) \div 2$ and $24 \div (4 \div 2)$

e $(14 - 6) - 3$ and $14 - (6 - 3)$

3 Put brackets in these calculations to make them correct.

a $7 + 2 \times 3 = 27$

b $12 \div 3 + 1 = 3$

c $11 - 4 - 2 = 9$

d $9 - 2 \times 3 - 1 = 20$

e $9 - 2 \times 3 - 1 = 5$

f $9 - 2 \times 3 - 1 = 14$

g $8 \times 3 - 2 = 12 - 2 \times 2$

4 Using all the numbers 1, 2, 3 and 4 once only, you can make answers of 1 and 2 as shown:

$12 \div (3 \times 4) = 1$

$2 \times 4 \div (3 + 1) = 2$

Find ways of making answers of 3, 4, 5, ... using all the numbers 1, 2, 3 and 4 once only.

How far can you go? Remember to use brackets where needed.

4^2 (four squared) = $4 \times 4 = 16$

2^3 (two cubed) = $2 \times 2 \times 2 = 8$

3^5 (three to the power of five) = $3 \times 3 \times 3 \times 3 \times 3 = 243$

1 Write the following using index notation.

a $4 \times 4 \times 4 =$ _____

b $2 \times 2 \times 2 \times 2 \times 2 =$ _____

c $3 \times 3 \times 3 \times 3 =$ _____

d $5 \times 5 \times 5 \times 5 \times 5 \times 5 =$ _____

2 Find the value of:

a $3^4 =$ _____

b $2^5 =$ _____

c $6^2 =$ _____

3 Write these in order of size, starting with the smallest:

3^3 2^5 5^2 4^2 2^6 _____

4 729 is unusual, because it is a square number and a cube number.

$27^2 = 9^3 = 729$

There is a number between 50 and 100 that is a square number and a cube number.

Can you find it?   _____

5 $5^2 = 25$, $12^2 = 144$ and $13^2 = 169$

$25 + 144 = 169$

We could write $5^2 + 12^2 = 13^2$

See if you can fill in the blanks:

$3^2 + 4^2 =$ ___^{2} ___$^2 + 8^2 = 10^2$

___$^2 + 24^2 = 25^2$ $12^2 + 16^2 =$ ___^{2}

6 Marsha has a box full of bricks. Each brick is a cube, and they are all the same size. They make a perfect cube, so the number of bricks is a cube number.

She knocks them over, and finds that she can make three different sized cubes using every brick.

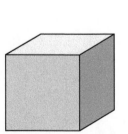
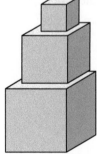

What is the smallest number of bricks Marsha could have started with?   _____

Roots

Student book topic 1.5

The square roots of 25 are the numbers whose square is 25. These are 5 and −5.

The square root sign refers to the positive square root, so $\sqrt{25} = 5$ and $-\sqrt{25} = -5$

The cube root of 64 is the number whose cube is 64.

This is $\sqrt[3]{64} = 4$

1 Without using a calculator, write down the value of:

a $\sqrt{49}$ = _____ **b** $\sqrt[3]{8}$ = _____ **c** $\sqrt{169}$ = _____ **d** $\sqrt[3]{125}$ = _____

2 Write down the solutions for these.

a $\sqrt{100}$ = _____ **b** $\sqrt[3]{1}$ = _____

c $\sqrt{196}$ = _____ **d** $\sqrt[3]{-1000}$ = _____

3 Use your calculator to find the values of these. Round your answers to 1 decimal place.

a $\sqrt{10}$ = _____ **b** $\sqrt[3]{51}$ = _____ **c** $\sqrt{150}$ = _____ **d** $\sqrt[3]{100}$ = _____

4 a A square has an area of 25 cm².

Find the length of each side. _____

b Another square has twice the area of this square.

Find the length of each side of the new square. _____

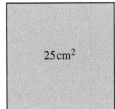

25 cm²

 5 A cube has a volume of 343 cm³.

Find the area of each face. _____

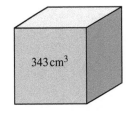

343 cm³

6 Calculate the value of these.

a i $\sqrt{49} - 5$ = _____ **ii** $-\sqrt{49} - 5$ = _____

b i $25 - \sqrt{25}$ = _____ **ii** $25 - (-\sqrt{25})$ = _____

c i $-\sqrt{16} - \sqrt[3]{8}$ = _____ **ii** $\sqrt{16} - \sqrt[3]{8}$ = _____

d i $\sqrt{400} + \sqrt[3]{1000}$ = _____ **ii** $-\sqrt{400} + \sqrt[3]{1000}$ = _____

Prime factors

> A prime number has exactly two factors, 1 and itself.
>
> A prime factor is a factor that is a prime number.

1 Complete the list of the prime numbers less than 20.

2, ———, ———, ———, 11, ———, ———, 19

2 Explain why 1 is not a prime number.

————————————————————————————————————

3 Complete the factor trees below.

a

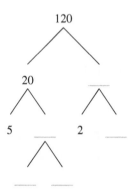

b

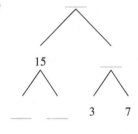

4 Write 250 as the product of its prime factors.

—————————————————

The highest common factor of two numbers is the product of all the prime factors shared by both numbers.

The lowest common multiple of two numbers is the product of all the prime factors in one or the other or both numbers.

1

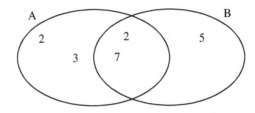

a Find the product of the 4 prime factors in oval A. _____

b Find the product of the 3 prime factors in oval B. _____

c Write down the highest common factor of your answers to **a** and **b**. _____

d What is the lowest common multiple of your answers to **a** and **b**? _____

2 Find the prime factors of 420 and 200 and complete the diagram.

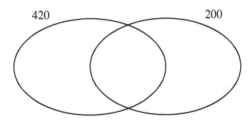

Highest common factor is _____

Lowest common multiple is _____

3 Three lights flash as follows.
Light A flashes every 9 seconds.
Light B flashes every 15 seconds.
Light C flashes every 12 seconds.
They all flash at the same time.
How long will it be before they all flash again at the same time?

_____

A number, x, is multiplied by 3. Then 2 is added. The answer is divided by 4. Then 6 is subtracted.

This can be written as $\dfrac{3x + 2}{4} - 6$

1 Match each statement with the correct expression.
The first one has been done for you.

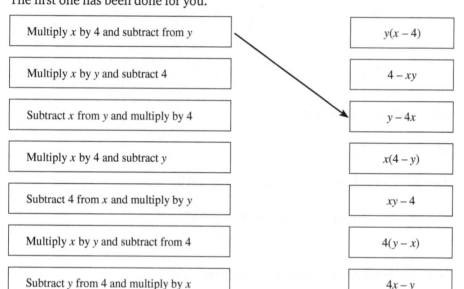

Multiply x by 4 and subtract from y	$y(x - 4)$
Multiply x by y and subtract 4	$4 - xy$
Subtract x from y and multiply by 4	$y - 4x$
Multiply x by 4 and subtract y	$x(4 - y)$
Subtract 4 from x and multiply by y	$xy - 4$
Multiply x by y and subtract from 4	$4(y - x)$
Subtract y from 4 and multiply by x	$4x - y$

2 Asif is x years old.
Write an expression to answer the following questions.

a How old will Asif be in 5 years' time? _____

b How old was Asif 3 years ago? _____

c Saeed is 5 years younger than Asif.
How old is Saeed? _____

d Amiira is 3 times as old as Saeed.
How old is Amiira? _____

e Saeed is twice as old as Karl.
How old is Karl? _____

3 A pencil costs m cents, a notebook costs n cents and a ruler costs r cents.
Write down an expression for the cost of 5 pencils, 4 notebooks and x rulers.

_____ cents

Multiplying algebraic terms

$x^2 \times x = x^3,\quad 3x \times 2xy = 6x^2y,\quad (3xy)^2 = 3xy \times 3xy = 9x^2y^2$

1 Match each expression on the top with one on the bottom which has the same answer. The first one has been done for you.

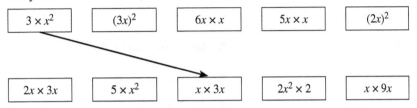

| $3 \times x^2$ | $(3x)^2$ | $6x \times x$ | $5x \times x$ | $(2x)^2$ |

| $2x \times 3x$ | $5 \times x^2$ | $x \times 3x$ | $2x^2 \times 2$ | $x \times 9x$ |

2 Fill in the gaps in these statements.

a $2x \times 3y = $ _____

b $3x^2 \times 4y = $ _____

c $2x \times xy = $ _____

d $3x \times $ _____ $= 6xy$

e _____ $\times 5x = 20x^2$

f $4a^2 \times $ _____ $= 4a^3b$

g $($_____$)^2 = 16x^2$

h $3ab \times $ _____ $= 12a^2b^3$

i $($_____$)^2 \times 2x = 18xy^2$

3 Complete the wall by filling in the gaps.
You multiply the terms in two blocks to find the term in the block above.

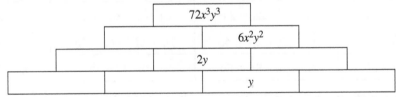

4 Complete these diagrams. On each line, the terms in the two circles multiply to give the term in the rectangle between them.

a

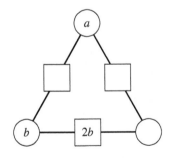

b

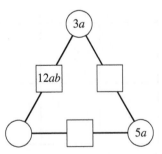

c

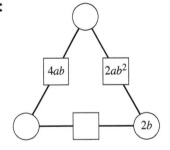

5 $2x \times 3xy = 6x^2y$

Find three different pairs of terms that multiply to $6x^2y$

_____ $\times$ _____ $= 6x^2y$ \qquad _____ $\times$ _____ $= 6x^2y$ \qquad _____ $\times$ _____ $= 6x^2y$

Collecting like terms

> Like terms contain exactly the same letters or combinations of letters. Like terms can be added and subtracted.
>
> $3a + 4b - 2a = a + 4b$ $6a^2 - 2a - 2a^2 + a = 4a^2 - a$

1 Simplify these expressions.

    a $5a + 3b - 3a =$ _____

    b $6a^2 - 3a - 2a^2 + 4a =$ _____

    c $5c - 2d^2 - 4c - 3d^2 =$ _____

    d $5t + 3st - t + 2st =$ _____

    e $a^2 + 4b^2 - 2a^2 + 3b =$ _____

    f $6xy - 3x - y - 2x + xy =$ _____

2 Complete these calculations.

    a $3x + 4y +$ _____ $-$ _____ $= 5x + 2y$

    b $2x^2 - 3x -$ _____ $-$ _____ $= x^2 - 5x$

    c $2xy +$ _____ $- xy - 2x =$ _____ $+ 3x$

    d $a^2 -$ _____ $+ 3a^2 + 4b =$ _____ $+ 2b$

 3 In each row, find the expression that is **not** equivalent to the other three.

$2x + 3y + x - 2y$	$4x + y - x$	$4y + 2x + x - 5y$	$5x + 6y - 2x - 5y$
$a^2 - 3a + 5a - 2a^2$	$4a - a^2 - 2a + 2a^2$	$2a^2 - a - 3a^2 + 3a$	$a + 2a^2 + a - 3a^2$
$4p - 2p^2 - 3p - p^2$	$p^2 + 5p - 2p^2 - 4p$	$7p + 3p^2 - 6p - 4p^2$	$2p - p - p^2$
$2a - 3b + a + 3b$	$5b - 2a - 5b + 5a$	$4b - 3a - 4b$	$4a - 3a^2 - a + 3a^2$

4 Complete this magic square so that each row, each column and both diagonals add up to $6x - 3y$

$5x - 4y$		$3x - 2y$
x		

To expand brackets, each term inside the brackets is multiplied by the term outside the brackets.

$3(a - 2b) - 2a(a - 3) = 3a - 6b - 2a^2 + 6a = 9a - 6b - 2a^2$

[Remember: $-2a \times -3 = +6a$]

1 Multiply out the following:

a $3(2x + 5) =$ _____

b $4(2y - 3) =$ _____

c $6(2a - 3b) =$ _____

d $a(5 + 2a) =$ _____

e $3b(2b - 4a) =$ _____

f $-2f(3f - 4) =$ _____

2 Multiply out and simplify:

a $3(2k - 4) + 4(2k - 1)$

$= 6k -$ _____

$=$ _____

b $5a(a - 3) - 2(a + 5)$

$= 5a^2 -$ _____

$=$ _____

3 Write down an expression for the area of the shape below.
Simplify your answer.

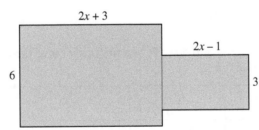

Area $= 6($ _____ $) +$ _____ $($ _____ $)$

$=$ _____ $+$ _____

$=$ _____

4 Simplify $3x(2x - 3y + 4) - 2y(3x - 2y - 1)$

$=$ _____

$=$ _____

$2a = 2 \times a, \frac{b}{3} = b \div 3$

BIDMAS = Brackets, Indices, Division and Multiplication, Addition and Subtraction

1 If $n = 4$, find the value of:

  a $2n + 3$ _____

  b $3n - 1$ _____

  c n^2 _____

  d $\frac{5n}{2}$ _____

  e $\frac{n^2}{8}$ _____

  f $n^2 - \frac{12}{n}$ _____

2 If $a = 3$ and $b = -2$, find the value of:

  a $3a + b$ _____

  b $3ab$ _____

  c $3(a + b)$ _____

  d ab^2 _____

  e $a + \frac{3b}{2}$ _____

  f $(ab)^2$ _____

3 When $c = 2$ and $d = -3$, which one of these expressions has a different value to all the others?

| $c - d$ | $4c + d$ | $1 - cd$ | $d^2 - 2c$ | $c^2 + 1$ |

4 If $r = -3$, $t = 2$ and $w = -1$, find the value of:

  a rtw _____

  b $t(r - w)$ _____

  c $r^2t - 3w$ _____

 5 Work through this maze when $a = 4$ and $b = -3$. You may only move through boxes, horizontally or vertically, which have a value of 1 or 5.

START	$\frac{b-5}{2}$	$\frac{3a+b}{3}$	$a^2 - 5b$	$b^2 + 2a$	$2a - b$
$2a + b$	$\frac{5a}{4}$	$\frac{2a}{4} + b$	$a + 1$	$\frac{3a-b}{3}$	$b + 8$
$b - 4$	$8a + b^3$	$ab - 13$	$b + 4$	$b - 2$	$b^2 - a$
$\frac{a}{2} - b$	$b^2 - 2a$	$b^2 + a$	$\frac{2a}{4} - b$	$a - b$	$a - \frac{b}{3}$
$a^2 + 5b$	$2a + \frac{b}{3}$	$8 - b$	$\frac{b^2}{9}$	$8a - b^3$	$a + b + 4$
$ab + 13$	$2 - b$	$\frac{b+5}{2}$	$a + b$	$ab - 7$	FINISH

Parallel lines

Vertically opposite angles are equal.

Alternate angles on parallel lines are equal.

Corresponding angles on parallel lines are equal.

1

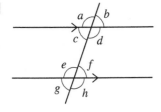

In the diagram above, complete these pairs of angles.

c and _____ are vertically opposite angles.

c and _____ are corresponding angles.

c and _____ are alternate angles.

e and _____ are vertically opposite angles.

d and _____ are corresponding angles.

e and _____ are alternate angles.

2 For each pair of angles, say whether they are vertically opposite, alternate or corresponding angles.

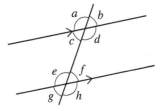

a and d are _____

b and f are _____

e and d are _____

f and g are _____

c and f are _____

h and d are _____

3 Calculate the angles marked with letters in this diagram.

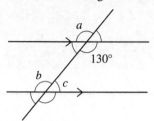

$a =$ _____ °

$b =$ _____ °

$c =$ _____ °

4

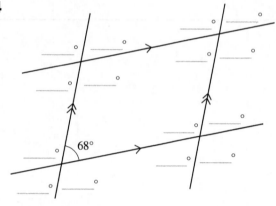

This diagram contains 16 angles.

One of them is marked 68°.

Use this to write in the sizes of the other 15 angles.

5

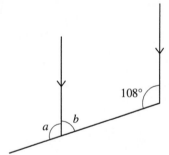

Calculate the size of the angles marked with letters, giving a reason in each case.

$a =$ _____ ° because _____ .

$b =$ _____ ° because _____ .

6 a

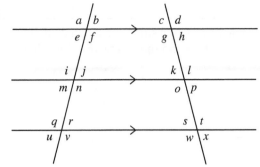

Some of these statements are true. Others are false.

After each statement write TRUE or FALSE.

$a = i$ _____ $m = j$ _____

$n = s$ _____ $e = n$ _____

$d = v$ _____ $w = d$ _____

$e = r$ _____ $o = r$ _____

b There are five angles that are equal to angle k.

Write down the name of each angle, giving a reason in each case

_____ because _____ .

_____ because _____ .

_____ because _____ .

_____ because _____ .

_____ because _____ .

The angles in a triangle add up to 180°. An isosceles triangle has 2 equal sides and 2 equal angles. An exterior angle is the angle outside the triangle between one side and the extension of another side.

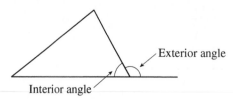

Exterior angle

Interior angle

1 Find the size of the angle marked *a* in this diagram.

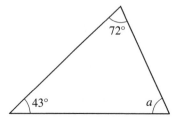

72°

43°

a

a = _____ °

2 Find the size of the angles marked *b* and *c* in this diagram.

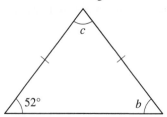

c

52°

b

b = _____ °

c = _____ °

3 Find the size of the angles marked *d* and *e* in this diagram.

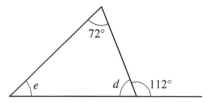

72°

e

d

112°

d = _____ °

e = _____ °

4 Calculate the size of each angle inside the triangle.

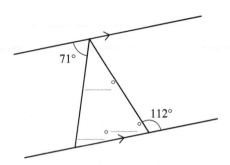

5 An isosceles triangle has an angle of 42°.
Find the size of the other two angles.
There are two possible solutions.

The other two angles are _____° and _____°

or the other two angles are _____° and _____°.

6 *ABC* is a triangle in which *AB* = *AC*
ABD is a straight line such that *CD* = *AB* = *AC*
Angle *ACB* = 72°

a Make a sketch of the shape in the space below.
Label the points *A, B, C* and *D*.

b What is the size of angle *ABC*? Give a reason for your answer.

ABC = _____° because _____.

c What is the size of angle *BAC*? Give a reason for your answer.

BAC = _____° because _____.

d What is the size of angle *ACD*? Give reasons for your answer.

ACD = _____° because _____.

and _____.

Congruent shapes

Congruent shapes are exactly the same shape and size.

1 Put a ring around any shapes that are congruent to shape *A*.

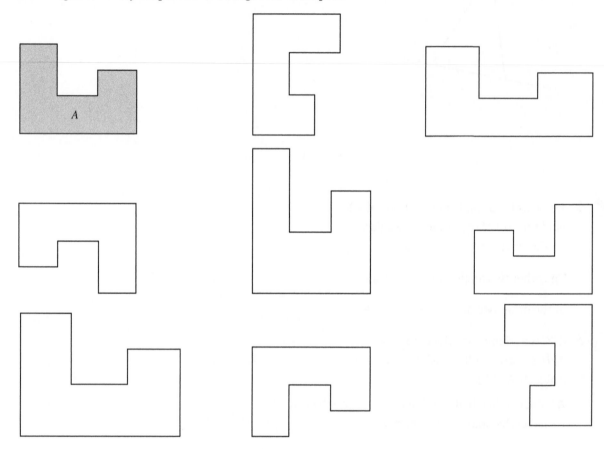

2 In these two congruent shapes, match the sides that correspond to each other.

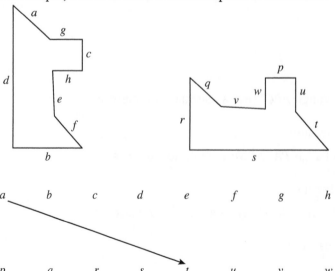

3 Split these shapes into two congruent parts. The first one is done for you.

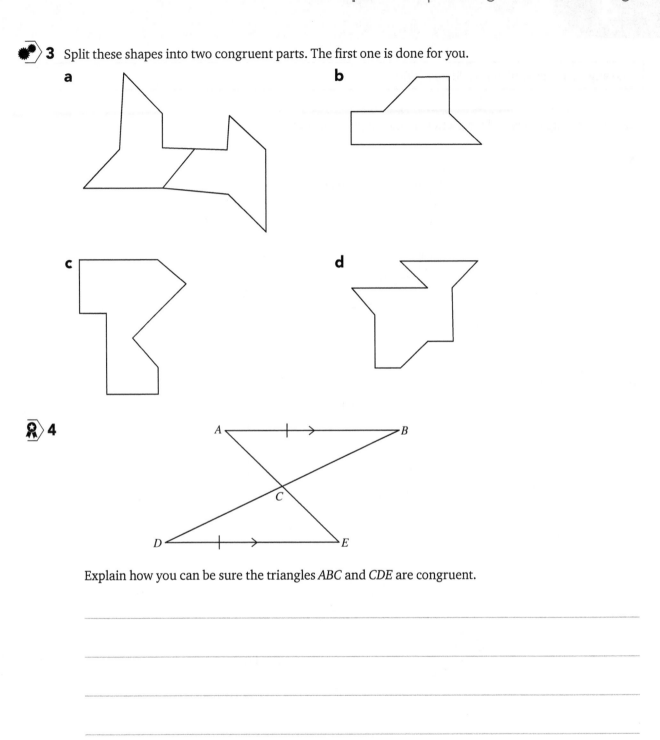

4

Explain how you can be sure the triangles *ABC* and *CDE* are congruent.

_____

_____

_____

_____

The angles in a quadrilateral add up to 360°.

Calculate the angles marked with letters in the diagrams below.

1

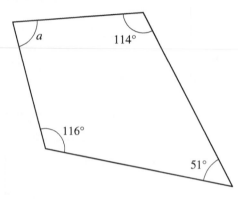

$a =$ _____ °

2

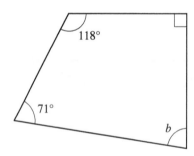

$b =$ _____ °

3

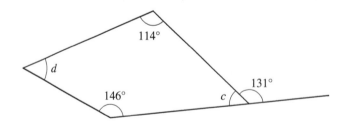

$c =$ _____ °

$d =$ _____ °

4

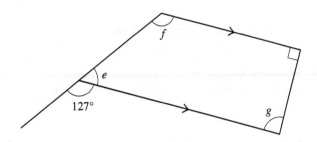

$e =$ _____ °

$f =$ _____ °

$g =$ _____ °

 5

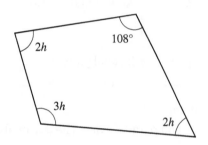

$h =$ _____ °

Complete the table. It will help with the questions that follow.

Shape	Sides equal	Sides parallel	Angles	Diagonals equal	Diagonals bisect	Diagonals cross at right angles
Square	All 4					
Rectangle	2 equal pairs		All 90°	Yes	Yes	
Parallelogram		2 pairs	Opposite angles equal			No
Rhombus						Yes
Kite			1 equal pair		No	
Trapezium (Not isosceles)	None necessarily equal	1 pair	None equal	No		
Isosceles trapezium		1 pair	2 equal pairs			No

1 Marlene is describing a quadrilateral.

She says, 'It has two pairs of equal and parallel sides. The diagonals are not equal.'

What shape is she describing?

_____

2 Which quadrilaterals have diagonals that are equal?

_____ , _____ and _____

3 Draw a kite with one right angle.

4 Draw a kite with two right angles.

5 Draw a trapezium with two equal sides that is not an isosceles trapezium.

6 Draw a trapezium with three equal sides.

7 For each of these descriptions, say whether they are possible or impossible.

   a A kite with one obtuse angle. _____

   b A kite with two obtuse angles. _____

   c A kite with three obtuse angles. _____

   d A trapezium with exactly two equal angles. _____

   e A trapezium with exactly three equal angles. _____

A net of a solid is a two-dimensional shape that can be folded to make the solid.

1 Which of these are nets of a cube? Put a ring around the correct ones.

2 When this net is folded to make a cube, the edges at *c* and *d* meet. Pair the other edges that meet.

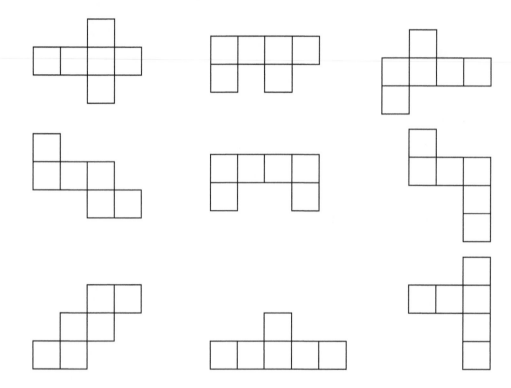

c and *d*   _____ and _____   _____ and _____

_____ and _____   _____ and _____

_____ and _____   _____ and _____

3 Draw an accurate net for this cuboid.

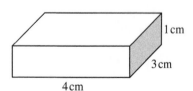

4 One of these nets will not make this square-based pyramid.
Which one? Put a ring around it.

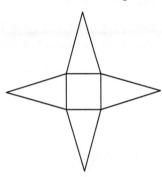

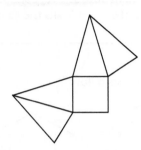

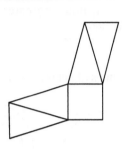

5 Make a drawing of the three-dimensional shape that this net will make.

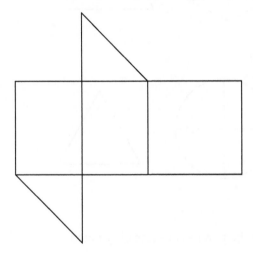

6 Draw a net of this shape.

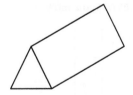

Symmetry

Student book topic 3.7

A shape has reflection symmetry if a mirror placed along the line of symmetry would reproduce the original shape.
Every shape looks the same when turned through 360° so every shape has rotation symmetry of order 1.
A shape has rotation symmetry greater than order 1 if it looks the same when rotated through an angle, or angles, less than 360°.

1 Draw in all the lines of symmetry on the shapes below.

2 Write down the order of rotation symmetry for these shapes.

 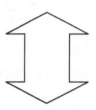 

_____   _____   _____

3 In each diagram below, shade in one more square so that the grid has one line of symmetry.

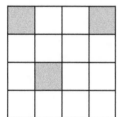   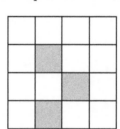   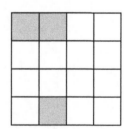

4 In each diagram below, shade in two more squares so that the grid has two lines of symmetry.

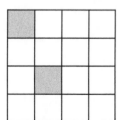   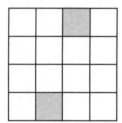   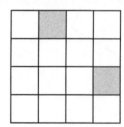

5 Write down the order of rotation symmetry of these shapes.

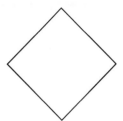  

_____   _____   _____

6 In each diagram, shade in two more squares so that the grid has the order of rotation symmetry given below.

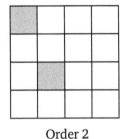 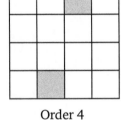 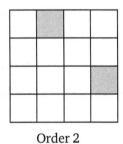

Order 2   Order 4   Order 2

7 Draw a quadrilateral with order of rotation symmetry 2 but no reflection symmetry.

8 Draw a quadrilateral with order of rotation symmetry 1 and one line of reflection symmetry.

9 a Is it possible to draw a shape with four lines of reflection symmetry and rotation symmetry of order 2?

_____

b Is it possible to draw a shape with rotation symmetry of order 4 and two lines of reflection symmetry?

_____

Comparing fractions

Student book topic 4.1

> Equivalent fractions have the same value.
>
> Multiply or divide the numerator and denominator by the same amount to find equivalent fractions.

1 Fill in the boxes.

a $\dfrac{2}{3} = \dfrac{\square}{6}$ **b** $\dfrac{3}{5} = \dfrac{9}{\square}$ **c** $\dfrac{3}{4} = \dfrac{\square}{20}$

d $\dfrac{7}{8} = \dfrac{\square}{48}$ **e** $\dfrac{5}{12} = \dfrac{30}{\square}$ **f** $\dfrac{2}{9} = \dfrac{\square}{72}$

2 Fill in the boxes.

a $\dfrac{12}{15} = \dfrac{\square}{5}$ **b** $\dfrac{24}{30} = \dfrac{4}{\square}$ **c** $\dfrac{30}{45} = \dfrac{\square}{3}$

d $\dfrac{36}{81} = \dfrac{\square}{9}$ **e** $\dfrac{50}{125} = \dfrac{2}{\square}$ **f** $\dfrac{12}{96} = \dfrac{\square}{8}$

3 Write these fractions in their lowest terms.

a $\dfrac{10}{15} =$ ____ **b** $\dfrac{8}{12} =$ ____ **c** $\dfrac{21}{24} =$ ____

d $\dfrac{28}{42} =$ ____ **e** $\dfrac{36}{63} =$ ____ **f** $\dfrac{40}{56} =$ ____

4 Explain how this diagram shows that $\frac{4}{5}$ is greater than $\frac{3}{4}$.

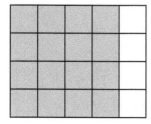 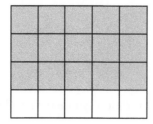

_____

_____

_____

5 Use equivalent fractions to compare these pairs of fractions.
Put a ring around the greater fraction in each pair.

a $\dfrac{5}{6}, \dfrac{2}{3}$ **b** $\dfrac{3}{4}, \dfrac{5}{8}$ **c** $\dfrac{3}{10}, \dfrac{2}{5}$

d $\dfrac{2}{3}, \dfrac{3}{5}$ **e** $\dfrac{1}{6}, \dfrac{3}{8}$ **f** $\dfrac{3}{14}, \dfrac{4}{21}$

Addition and subtraction of fractions and mixed numbers

Student book topic 4.2

Fractions need to have the same denominator in order to be added or subtracted.

1 Complete these.

a $\frac{3}{8} + \frac{1}{8}$

$= \frac{\square}{8}$

$= \frac{\square}{\square}$

b $\frac{2}{3} + \frac{1}{6}$

$= \frac{\square}{6} + \frac{\square}{6}$

$= \frac{\square}{\square}$

c $1\frac{2}{5} + 3\frac{3}{10}$

$= \underline{\quad} \frac{\square}{10} + \frac{\square}{10}$

$= \underline{\quad} \frac{\square}{\square}$

2 Work these out, simplifying your answer where possible.

a $\frac{5}{12} - \frac{1}{12}$

b $\frac{4}{9} - \frac{1}{3}$

c $3\frac{5}{8} - 1\frac{1}{4}$

3 Work these out, simplifying your answer where possible.

a $\frac{1}{4} + \frac{2}{3}$

b $\frac{4}{5} - \frac{1}{2}$

c $4\frac{5}{6} - 2\frac{1}{4}$

4 Work these out, simplifying your answer where possible.

a $2\frac{3}{4} + 1\frac{2}{5}$

b $3\frac{1}{5} - 1\frac{1}{3}$

c $4\frac{2}{3} - 1\frac{3}{4}$

5 Complete these walls so that the numbers in two bricks add to the number in the brick above. Give your answers in their lowest terms.

a

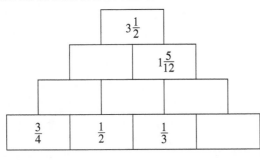

b

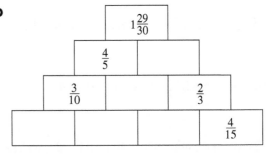

Multiplying and dividing an integer by a fraction

To find a fraction of a quantity, multiply the fraction by the quantity.

To multiply an integer by a fraction, multiply the numerator by the integer.

To divide an integer by a fraction, turn the fraction upside down and multiply it by the integer.

1 Complete these.

a $3 \times \frac{2}{7} =$ ——

b $5 \times \frac{1}{8} =$ ——

c $2 \times \frac{4}{9} =$ ——

2 Work out these, giving answers as mixed numbers in their simplest form.

a $3 \times \frac{5}{9}$

b $6 \times \frac{3}{8}$

c $4 \times \frac{3}{5}$

3 Work out these, giving answers as mixed numbers in their simplest form.

a $3 \div \frac{2}{3}$

b $4 \div \frac{2}{5}$

c $9 \div \frac{3}{8}$

4 Find $\frac{3}{5}$ of 8.

5 $\frac{3}{5}$ of a number is 12. What is the number?

6 How much bigger is $4 \div \frac{2}{3}$ than $4 \times \frac{2}{3}$?

7 What number, when divided by $\frac{3}{5}$, gives an answer of $6\frac{2}{3}$?

Integer powers of 10

Student book topic 5.1

> To multiply by $\frac{1}{10}$ or 0.1, you divide by 10.
>
> To divide by $\frac{1}{10}$ or 0.1, you multiply by 10.

1 Match the numbers in figures with the numbers in words.

1 230 000	Twelve thousand, three hundred
12 300	One thousand, two hundred and thirty
123 000	One hundred and twenty-three million
1230	One million, two hundred and thirty thousand
12 300 000	One hundred and twenty-three thousand
123 000 000	Twelve million, three hundred thousand

2 Write these numbers in words.

a 2450 _____

b 35 000 _____

c 4 000 400 _____

d 10 101 010 _____

_____

3 Fill in the gaps with one of these: × 0.1   × 0.01   ÷ 0.1   ÷ 0.01

a 4.2 _____ = 0.042

b 1.25 _____ = 125

c 0.7 _____ = 70

d 12 _____ = 1.2

4 Work out

    a $1.6 \times 0.1 =$ _____

    b $3.4 \div 0.01 =$ _____

    c $12.3 \times 0.01 =$ _____

    d $1.34 \div 0.1 =$ _____

    e $0.56 \div 0.01 =$ _____

    f $0.54 \times 0.01 =$ _____

5 Fill in the gaps in these questions and answers.

    a $4.1 \times 0.1 =$ _____

    b _____ $\div 0.01 = 27.3$

    c _____ $\times 0.01 = 0.46$

    d _____ $\div 0.1 = 1.4$

    e _____ $\div 0.01 = 34.6$

    f _____ $\times 0.01 = 0.004$

Rounding

When rounding, look at the next figure to decide whether or not to round up.

If a number is halfway between two whole numbers, you usually round up.

1 Put a ring around the numbers below that round to 60, when rounded off to the nearest 10.

58   65   54.9   64.9   6.29   62

2 Round these numbers:

a 24.67 to one decimal place _____

b 5467 to the nearest hundred _____

c 1.2345 to two decimal places _____

d 6.78 to the nearest ten _____

3 David gives the answer to a question as 56.

This is correct to the nearest whole number.

What is the smallest number the answer could have been before rounding?

_____

4 Complete the table below.

Number	To the nearest 10	To the nearest 100	To 1 decimal place	To 2 decimal places
234.567	230		234.6	
162.47				
2345.67				
78.123				
9.41467				

5 Fatima rounds 3.96 to 1 decimal place and gives the answer as 4.

Explain why Fatima is wrong.

_____

_____

6 The attendance at a football match was reported to be 23 600 to the nearest hundred.

   a What was the smallest possible attendance if the report was true?

   _____

   b What was the greatest possible attendance if the report was true?

   _____

7 Match the statements on the left with the numbers on the right.

23.456 to 1 decimal place
2.39 to 1 decimal place
23.96 to 1 decimal place
23.391 to 1 decimal place
23.9 to the nearest whole number
23.349 to 1 decimal place
25.23 to the nearest whole one

2.4
24
23.5
25
23.3
23.4
24.0

Calculating with decimals

Align the decimal points to add or subtract decimals.

When multiplying by a decimal, position the decimal point after the calculation.

When dividing by a decimal, use an equivalent calculation so you divide by an integer.

1 Work these out.

a 56.4
+24.7

b 37.8
−19.2

c 14.52
+37.67

d 9.165
−4.6

e 11.23
−6.563

2 Work out

a $7 \times 0.4 =$ _____

b $0.8 \times 9 =$ _____

c $6 \times 0.03 =$ _____

d $0.02 \times 80 =$ _____

e $0.6 \times 0.3 =$ _____

f $1.2 \times 0.6 =$ _____

3 Work out

a 7.4×0.4

b 0.06×3.4

c 24×0.04

d $42 \div 0.6$

e $61.5 \div 5$

f $3.84 \div 0.8$

g $3.2 \div 0.04$

h $0.048 \div 0.06$

4 Work these out. Give your answers to 2 decimal places.

a $5.83 \div 0.6$

b $2.67 \div 0.8$

c $3 \div 0.7$

5 Use the fact that $27 \times 34 = 918$ to write down the answers to

a 2.7×3.4

b $91.8 \div 270$

c $9.18 \div 3.4$

d 54×34

Fractions and decimals

To change a fraction to a decimal, divide the numerator by the denominator.

1 $36 > 4$

Explain why $0.36 < 0.4$

_____

_____

2 Write each set of numbers in order of size, starting with the smallest.

a 2.4   2.51   2.38   3.45   3

_____   _____   _____   _____   _____

b 0.78   0.758   0.8   0.7485   0.76

_____   _____   _____   _____   _____

c 11   10.78   11.23   10.6789   11.155

_____   _____   _____   _____   _____

d 1.07   1.701   7   1.1771   1.71

_____   _____   _____   _____   _____

3 Match these fractions with the correct decimal.

$\frac{2}{5}$	$\frac{1}{4}$	$\frac{3}{8}$	$\frac{3}{10}$	$\frac{7}{20}$	$\frac{11}{40}$

0.375	0.275	0.4	0.35	0.3	0.25

4 Match these fractions with the correct decimal.

$\frac{2}{7}$	$\frac{7}{24}$	$\frac{3}{11}$	$\frac{3}{13}$	$\frac{2}{9}$	$\frac{4}{15}$

$0.\dot{2}\dot{7}$	$0.\dot{2}8571\dot{4}$	$0.\dot{2}$	$0.2\dot{6}$	$0.\dot{2}3076\dot{9}$	$0.291\dot{6}$

5 Change these fractions to recurring decimals.

$$\frac{4}{9}$$ $$\frac{4}{7}$$ $$\frac{6}{11}$$

_____   _____   _____

6 Change these fractions to decimals.

Give your answers to 2 decimal places.

$$\frac{5}{12}$$ $$\frac{13}{15}$$ $$\frac{7}{24}$$

_____   _____   _____

7 Write these fractions in order, starting with the smallest.

Change them to decimals to help.

$$\frac{5}{7},$$ $$\frac{4}{5},$$ $$\frac{8}{11},$$ $$\frac{18}{25},$$ $$\frac{17}{24}$$

_____ , _____ , _____ , _____ , _____

8 a Write $\frac{3}{11}$ as a decimal. _____

b Write as $\frac{8}{11}$ a decimal. _____

c Add the fractions $\frac{3}{11}$ and $\frac{8}{11}$ _____

d Add together your decimal answers from **a** and **b**. _____

e What do you notice about your answers to **c** and **d**?

_____

_____

Exercise 1 · Calculating statistics from data sets

Student book topic 6.1

Chapter 18 covers collecting data
Chapter 10 covers presenting data

> The mean is $\dfrac{\text{the sum of the values}}{\text{the number of values}}$
>
> The median is the middle value when they are put in order.
>
> The mode is the most common value.
>
> The range is the difference between the largest value and the smallest value.

1 Find the mean of 3, 6, 7, 9, 10, 13.

_____

2 Find the median of 5, 10, 6, 19, 12, 8.

_____

3 Find the mode of 6, 12, 5, 7, 6, 3, 10, 8.

_____

4 Find the range of 2, 6, 1, 8, 5, 4, 10.

_____

5 Five numbers have a mean of 7, a mode of 4, a median of 8 and a range of 6.

Find the five numbers.

6 Ravi carried out a survey to find the number of pets belonging to the children in his class.

Here are his results.

Number of pets	Frequency
0	2
1	7
2	5
3	5
4	3
5	2
6	1

a Find the median number of pets in the class. _____

b Find the mean number of pets in the class. _____

c Find the modal number of pets in the class. _____

d Find the range of the number of pets in the class. _____

e Another class carries out the same survey with these results:

Number of pets	Frequency
0	3
1	6
2	3
3	5
4	5
5	1
6	2

Find the mean number of pets when the two sets of data are combined. _____

When data are grouped, assume all the data are equal to the middle value of each group.

1 The table below shows the number of days absence from school in a year for a group of children.

Number of days absence	Frequency
0–4	11
5–9	8
10–14	4
15–19	0
20–24	1

a Find the class containing the median.

b Find the modal class.

c Calculate an estimate for the mean number of days absence.

d Explain why your answer to part **c** is only an estimate.

2 The table below shows the heights of some tomato plants.

Height, h cm	Frequency
$15 \leqslant h < 20$	1
$20 \leqslant h < 25$	5
$25 \leqslant h < 30$	11
$30 \leqslant h < 35$	10
$35 \leqslant h < 40$	3

a Calculate an estimate for the mean height.

_____

b Two weeks later, the heights of the tomato plants were as shown below:

Height, h cm	Frequency
$20 \leqslant h < 25$	1
$25 \leqslant h < 30$	5
$30 \leqslant h < 35$	9
$35 \leqslant h < 40$	11
$40 \leqslant h < 45$	4

Calculate an estimate for the mean increase in height during the two weeks.

_____

Making comparisons

To compare two sets of data, use the range and at least one of mean, median and mode.

1 The tables show the number of peas in pods for two different varieties of pea.

Type A	
Number of peas	**Frequency**
3	2
4	15
5	12
6	9
7	6
8	1

Type B	
Number of peas	**Frequency**
3	0
4	11
5	14
6	10
7	4
8	1

a Calculate the mean for each type.

Type A mean _____   Type B mean _____

b Find the range for each type.

Type A range _____   Type B range _____

c Which type seems to be better? Give a reason for your answer.

Type _____ appears better because _____

_____

2 Nadia uses batteries in her torch.

She has ten type A batteries, and ten type B batteries.

She records how long they last.

Here are her results.

Lifetime (hours)										
Type A	27	28	27	29	26	29	28	30	26	28
Type B	24	26	40	38	25	25	27	26	25	26

a Find the mean lifetime of each type and the median lifetime of each type.

Type A mean = _____ hours Type B mean = _____ hours

Type A median = _____ hours Type B median = _____ hours

b Say which type of batteries Nadia should use in the future, and say why.

Nadia should use type _____ batteries because _____

_____

3 Andrew, Hassan, Ramesh and Sunil play cricket for the same club.

Here are their scores in their last 6 innings

Andrew	45	48	23	8	23	109
Hassan	23	43	56	74	23	12
Ramesh	33	56	23	127	6	11
Sunil	22	55	65	11	44	24

a Which two cricketers have the same mode?

_____ and _____

b Which two cricketers have the same median?

_____ and _____

c Which two cricketers have the same mean?

_____ and _____

d Who do you think is the best batsman?

Give a reason for your answer.

I think _____ is the best batsman because _____

_____

The metric system

Student book topic 7.1

kilo = 1000, centi = $\frac{1}{100}$, milli = $\frac{1}{1000}$

1 Match the measurements with the correct units.

The distance between two towns	
The mass of a chair	
The amount of water in a bath	
The height of a house	
The capacity of a teaspoon	
The length of a worm	
The mass of a pencil	

g
ml
km
kg
l
cm
m

2 A bottle of drink has a mass of 840 g.

A bag can safely hold a mass of 5 kg.

What is the greatest number of bottles that can be carried safely in the bag?

_____ bottles

3 Change these measurements into the given units.

a 4 m = _____ cm

b 1.6 km = _____ m

c 1400 ml = _____ l

d 4500 g = _____ kg

e 300 g = _____ mg

f 45 cl = _____ l

g 2.4 l = _____ ml

h 12.4 tonnes = _____ kg

4 A drinking glass can hold 300 ml of drink.

How many glasses can be filled completely from a 2 litre jug?

_____ glasses

5 A drink is made from this recipe.

45 cl strawberry juice
20 cl watermelon juice
2 freshly squeezed oranges
1 freshly squeezed lime

An orange contains about 40 ml of juice.
A lime contains about 25 ml of juice.
How many millilitres of drink will be made altogether?

_____ ml

6 Ben makes a wooden picture frame.
He uses a piece of wood 3 m long.
He cuts two lengths of 85 cm each and two lengths of 45 cm each.
Find the length of the wood that is left over.
Give your answer in metres.

_____ m

7 Pablo cycles to work and back on five days per week.
The distance from his home to work is 2.3 km.
How far does Pablo cycle in 4 weeks?

_____ km

> Area is measured in square units (mm², cm², m², km²).
>
> Volume is measured in cubic units (mm³, cm³, m³, km³).

1 Match the measurements to the most appropriate units.

The area of the Pacific Ocean		cm³	
The volume of a DVD case		cm²	
The area of an envelope		m³	
The area of the head of a pin		km²	
The area of a tennis court		mm²	
The volume of a swimming pool		km³	
The volume of the Sun		m²	

2 Find the area and perimeter of these shapes. Give the correct units for your answers.

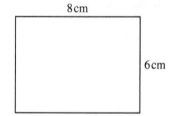

Area = _____   Area = _____

Perimeter = _____   Perimeter = _____

3 a Sarah is filling boxes with grass seed.
Each box is 12 cm long, 6 cm wide and 15 cm tall.
She has 5000 cm³ of grass seed.

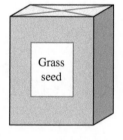

Grass seed

  i How many boxes can she fill completely?

_____ boxes

  ii How much grass seed is left over?

_____ cm³

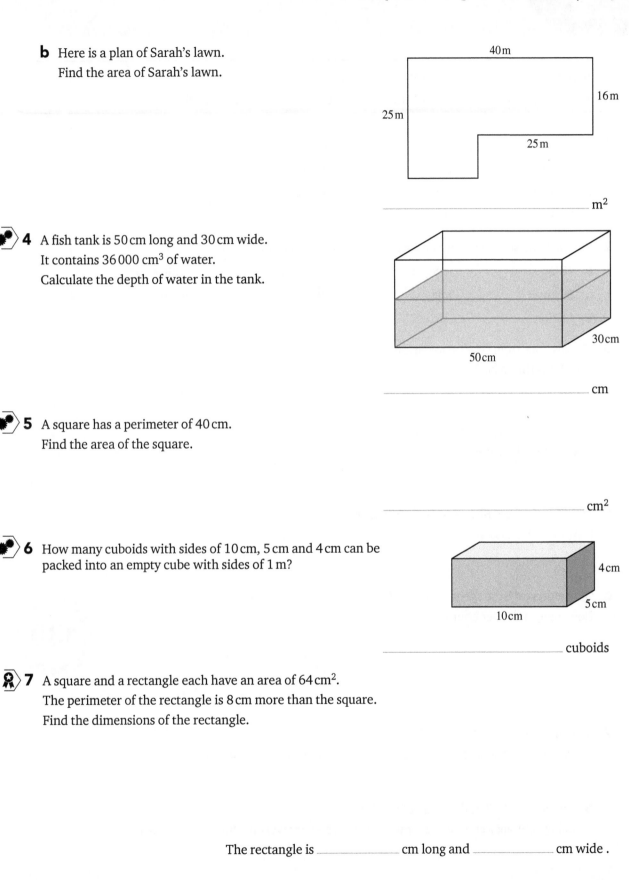

b Here is a plan of Sarah's lawn.
Find the area of Sarah's lawn.

40m

16m

25m

25m

_____ m²

4 A fish tank is 50 cm long and 30 cm wide.
It contains 36 000 cm³ of water.
Calculate the depth of water in the tank.

30cm

50cm

_____ cm

5 A square has a perimeter of 40 cm.
Find the area of the square.

_____ cm²

6 How many cuboids with sides of 10 cm, 5 cm and 4 cm can be
packed into an empty cube with sides of 1 m?

4cm

5cm

10cm

_____ cuboids

7 A square and a rectangle each have an area of 64 cm².
The perimeter of the rectangle is 8 cm more than the square.
Find the dimensions of the rectangle.

The rectangle is _____ cm long and _____ cm wide .

Non-metric units

5 miles ≈ 8 km	1 inch ≈ 2.5 cm	1 foot ≈ 30 cm
1 gallon ≈ 4.5 litres	1 litre ≈ 1.75 pints	1 kg ≈ 2.2 lb

1 Complete these, giving your answers to the nearest whole one.

   a 20 miles ≈ _____ km   **b** 67 miles ≈ _____ km

   c 44 miles ≈ _____ km   **d** 38 miles ≈ _____ km

   e 3 miles ≈ _____ km

2 Complete these, giving your answers to the nearest whole one.

   a 40 km ≈ _____ miles   **b** 10 km ≈ _____ miles

   c 20 km ≈ _____ miles   **d** 7 km ≈ _____ miles

   e 147 km ≈ _____ miles

3 The distance from London to Paris is about 285 miles.
How far is this in km?

_____ km

4 Tara drives 12 000 miles in a year.
How many kilometres is this?

_____ km

5 The distance from Bhopal to Delhi is 752 km.
How many miles is this?

_____ miles

6 On a motorway, the speed limit is 130 km/h.
How many miles per hour is this?

130

_____ mph

7 a Pierre's car holds 40 litres of diesel.
How many gallons is that? Give your answer to the nearest whole one.

_____ gallons

   b His car travels 40 miles on a gallon of diesel.
Using your answer to part **a**, how many km can he travel on a full tank of diesel?

_____ km

Solving equations using flow diagrams

You must perform the opposite operation and reverse the order.

Fill in the gaps to solve these equations.

1 $3x - 4 = 11$

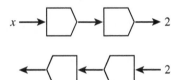

2 $5x + 4 = 39$

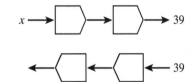

3 $\dfrac{x + 3}{4} = 2$

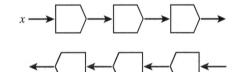

4 $\dfrac{2x - 3}{5} = 3$

Draw your own flow diagrams to solve these equations.

5 $7x - 5 = 2$

6 $\dfrac{x + 3}{2} = 7$

7 $\dfrac{2x - 1}{3} = 11$

8 $\dfrac{2x - 5}{3} + 9 = 20$

Solving equations with the unknown on one side

Student book topic 8.2

Remove any brackets first.

Always do the same thing to both sides of the equation.

1 Complete the following solution by filling in the gaps.

$$4x - 5 = 17$$

$$4x - 5 + 5 = 17 \underline{\hspace{1cm}}$$

$$4x = \underline{\hspace{1cm}}$$

$$x = \frac{\square}{4}$$

$$x = \underline{\hspace{1cm}}$$

2 Complete this equation by filling in the gaps.

$$5x - 3 = 32$$

$$5x = \underline{\hspace{1cm}}$$

$$x = \underline{\hspace{1cm}}$$

3 Solve these equations showing your working out.

a $7x - 2 = 19$

b $2(x + 1) = 11$

c $15 - 3x = 9$

d $\frac{x}{4} + 2 = 10$

e $4(x + 2) - 2(x - 5) = 8$

 4 This shed is 250 cm wide.

It has two windows, each x cm wide.

The gaps between the walls and the windows and between the windows are all 40 cm.

Find the value of x.

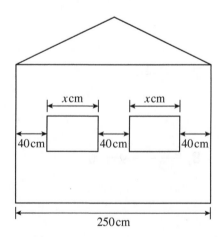

$x = \underline{\hspace{1cm}}$

Solving equations with the unknown on both sides

Collect the letters on one side of the equation and the numbers on the other side.

1 Complete the following equation by filling in the gaps.

$5x - 7 = 2x + 5$

_____ $x - 7 = 5$ (Subtract $2x$ from both sides)

_____ $x =$ _____ (Add 7 to both sides)

$x =$ _____ (Divide both sides by 3)

2 Solve these equations.

a $3x + 5 = 11 - x$

b $12x - 48 = 3x - 3$

$x =$ _____

$x =$ _____

c $3(2x - 1) = 4x + 7$

$x =$ _____

3 Solve these equations.

a $5x - 2 = 3x + 9$

b $4x + 7 = 2x - 3$

$x =$ _____

$x =$ _____

c $5(x - 3) = 2(2x + 6)$

$x =$ _____

 4 $6x - 2 = $ ⬛$x + 13$

The answer to the equation is $x = 5$.

What number is hidden by the ink blot?

_____ is hidden.

> You need to write an equation showing that two things are equal, and then solve the equation.

Answer all these problems by writing an equation and using the balance method to solve it.

1 Luca thinks of a number.
He multiplies it by 7 and then subtracts 5.
The answer is 51.
What was the number that Luca thought of?

Luca thought of the number _____

2 Three consecutive numbers add up to 96.
What are the three numbers?

The numbers are _____, _____ and _____.

3 The square and the triangle have the same perimeter.
Find the value of x.

$x =$ _____

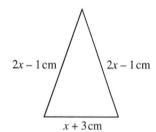

4 Shiraz is x years old.
Shiraz's dad is 24 years older than Shiraz.
Next year, Shiraz's dad will be four times as old as Shiraz.
How old is Shiraz?

Shiraz is _____ years old.

5 Eva has a large collection of books.
She has read exactly $\frac{3}{4}$ of her books.
She has read $5x + 1$ of them but she still has $2x - 15$ to read.
How many books does she have altogether?

Eva has _____ books.

6 In this wall, you add two blocks to find the number in the block above.
Find the value of x.

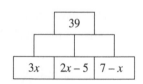

$x =$ _____

Constructions

> To construct the perpendicular bisector of a line, draw arcs from each end of the line and join the intersections of the arcs.
>
> To bisect an angle, draw equal arcs along each arm and then draw equal arcs from these intersections.

1 Complete this construction of the bisector of angle *ABC*.

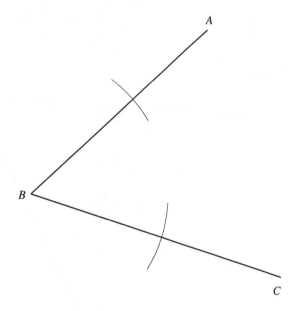

2 Complete this construction of the perpendicular bisector of *PQ*.

3 Construct the perpendicular bisector of *FG*.

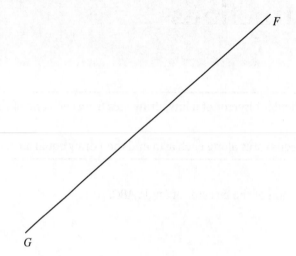

4 Construct the bisector of angle *JKL*.

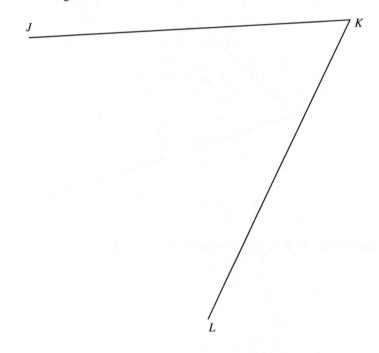

5 a Construct the perpendicular bisector of *AB*.

   b Bisect angle *BAC*.

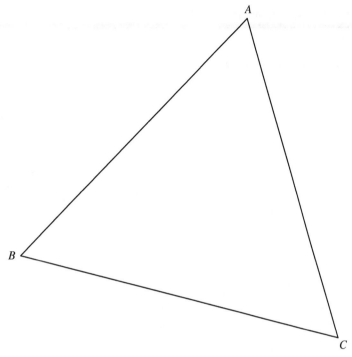

6 In a triangle *XYZ*, the perpendicular bisector of *XY* and the bisector of angle *XZY* are the same line. What does this tell you about triangle *XYZ*?

Draw a sketch if it helps.

To construct a triangle given three sides, draw one side and use compasses to find the third vertex.

To construct a right angle, bisect an angle of 180°.

1 Complete the construction of this triangle *ABC*.

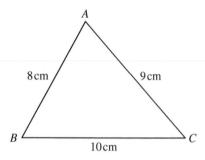

B _____ 10 cm _____ C

2 Construct a right angle at *P*.

P

3 a Construct triangle *XYZ*.

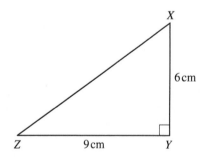

b Measure the length of *XZ*.

XZ = _____ cm

4 Construct a triangle with sides of 10 cm, 7 cm and 6 cm.

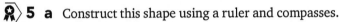

5 a Construct this shape using a ruler and compasses.

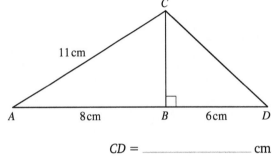

$CD =$ _____ **cm**

b Bisect angle *CAB*.

c Measure the length of *CD*.

When drawing circles and arcs, make sure the compasses are tight and use a sharp pencil.

1 Draw a circle with a radius of 3.5 cm, with the centre at *O*.

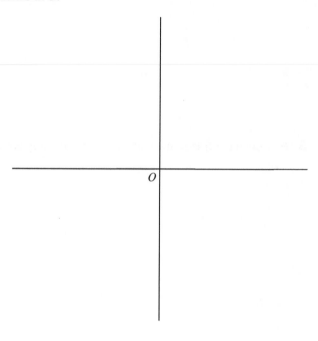

2 Make an accurate drawing of this shape.

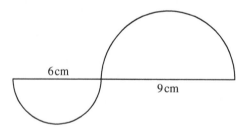

6 cm

9 cm

3 Make an accurate drawing of this shape, using semicircles with radii of 3 cm, 4 cm, 5 cm and 6 cm.

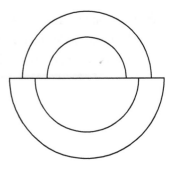

4 Make a copy of this shape, using semicircles with radii 1 cm, 2 cm, 3 cm, 4 cm and 5 cm.

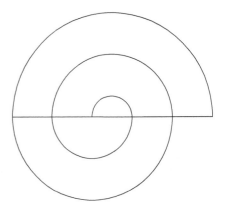

5 Make an accurate drawing of this shape.
The large circle and quarter circles have radii of 6 cm.
The small circle has a radius of 3 cm.

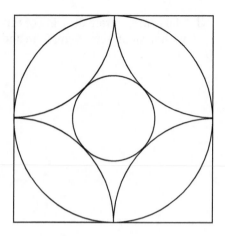

Make sure you look closely at the scale when making or reading from a scale drawing.

1 The scale drawing shows a bungalow.
The scale is 1 cm to 1 m.
Give all answers to 1 decimal place.

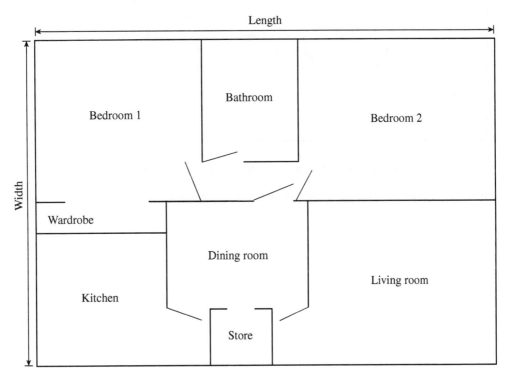

a What is the length of the real bungalow?

_____ m

b What is the width of the real bungalow?

_____ m

c What is the length and width of bedroom 1, excluding the wardrobe?

Length = _____ m, width = _____ m

d There is a bed, 2 m long and 1.5 m wide, in bedroom 2.

Draw this bed to the correct scale in bedroom 2.

2 Here is a rough sketch of a rectangular garden.

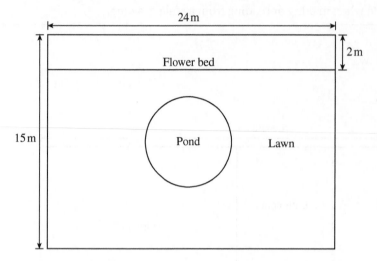

The pond has a diameter of 8 m, and is in the exact centre of the garden.

Use a scale of 1 cm to 2 m to make a scale drawing of the garden.

3 Here is a sketch of half a football pitch. It is 85 m wide.

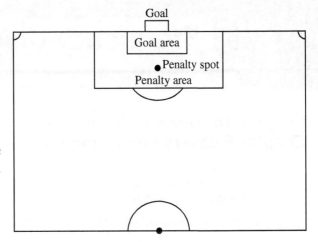

The length of the half pitch is 55 m.

The goal is 7.3 m wide.

The goal area is 18.3 m wide and 5.5 m long.

The penalty area is 40.3 m wide and 16.5 m long.

The penalty spot is 11 m from the goal line.

The arcs on the penalty area and the centre circle have a radius of 9.15 m.

The arcs at the corners have a radius of 1 m.

Make a scale drawing of the half pitch, using a scale of 1 cm to 5 m.

Frequency diagrams for grouped data

Student book topic 10.1

Chapter 18 covers collecting data
Chapter 6 covers processing data

Frequency diagrams have a title, labelled axes, evenly spaced values on the frequency axis (starting at zero) and bars of equal width.

For discrete data, use equally sized gaps between the bars.

For continuous data, there should be no gaps between the bars.

1 The frequency diagram shows the scores of a group of children in a maths test.

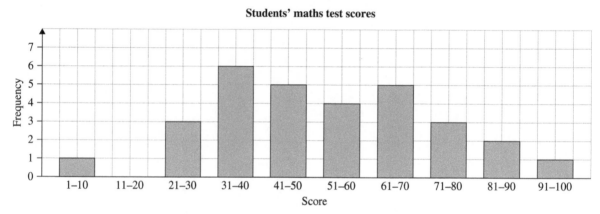

Students' maths test scores

a How many students scored between 31 and 40?

_____

b How many students scored 30 or less?

_____

c How many students took the test altogether?

_____

d What percentage of the students scored over 70?

_____

2 The table below shows how much sport is played by a group of children each week.

Number of hours, h	$0 \leq h < 1$	$1 \leq h < 2$	$2 \leq h < 3$	$3 \leq h < 4$	$4 \leq h < 5$	$5 \leq h < 6$
Frequency	4	9	6	4	0	2

Show this information in a frequency diagram.

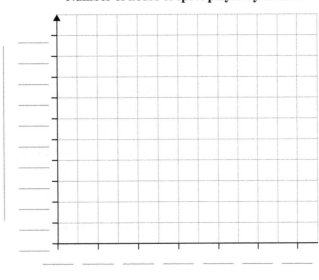

Number of hours of sport played by children

3 Find three things wrong with this frequency diagram showing the heights of some sunflowers in a field.

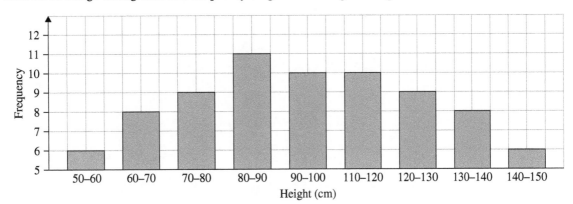

a _____

b _____

c _____

4 a Draw a frequency diagram to show this information about the ages of people on a bus.

Age (years)	1–10	11–20	21–30	31–40	41–50	51–60	61–70	71–80
Frequency	2	5	7	6	11	19	12	3

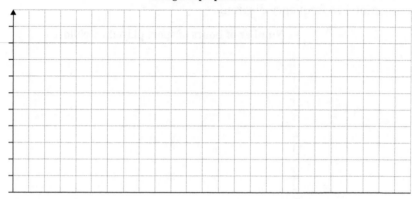

The ages of people on a bus

b Which is the modal age group?

_____

c What percentage of the passengers are aged between 21 and 40?

_____

The angles in a pie chart add up to 360°.

Divide 360° by the total to get the angle for a single item.

1 The pie chart shows the colour of cars passing a school one morning.

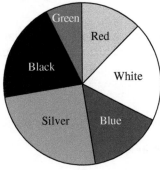

a Which colour was the least popular?

_____

b There were 10 silver cars.
How many cars passed the school altogether?

_____

c How many of the cars were blue?

_____

2 24 children and 36 adults were asked to choose their favourite fruit from a list.
The results are shown in the pie charts.

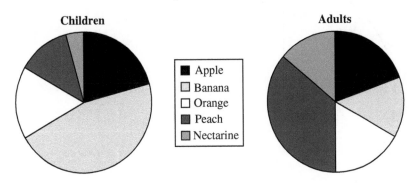

a How many children chose peach?

_____

b Which two fruits were chosen by the same number of adults?

_____

c How many more adults than children chose apple?

_____

d Both pie charts have the same size sector for orange. Explain how you know that more adults than children chose orange. You do not need to do any calculations.

_____

3 The chart below shows the favourite sports of people in a survey.

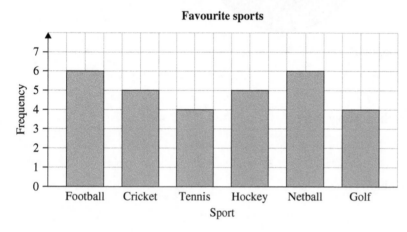

Show this information in a pie chart.

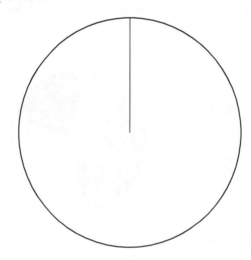

Always put time on the horizontal axis.

Line graphs show how data changes, or trends.

1 The temperature in Kamini's garden was recorded every hour during one day.

The graph shows her results.

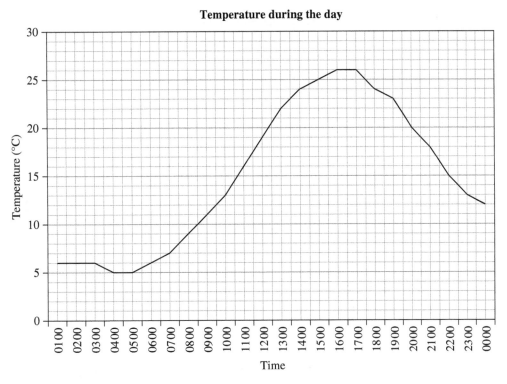

Temperature during the day

a What was the temperature in the garden at 15 00?

_____ °C

b By how much did the temperature fall between 17 00 and 20 00?

_____ °C

c Estimate the temperature at 13 30.

_____ °C

2 The table shows the average monthly rainfall in Rome.

Month	Jan	Feb	Mar	Apr	May	Jun	Jul	Aug	Sep	Oct	Nov	Dec
Average rainfall (mm)	25	21	19	28	18	5	7	16	26	38	49	38

a Show this information on the graph below.

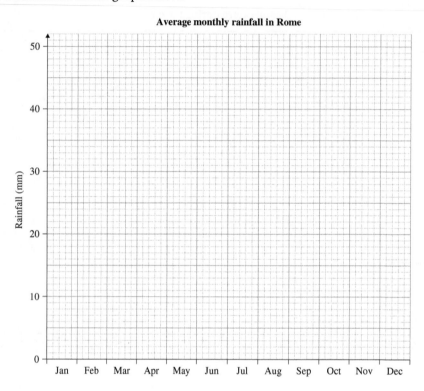

Average monthly rainfall in Rome

b For how many months is the average rainfall less than 20 mm?

_____ months

c Which two months have the same average rainfall?

_____ and _____

3 The graph below shows how many minutes Bruno and Chris spent playing computer games each day.

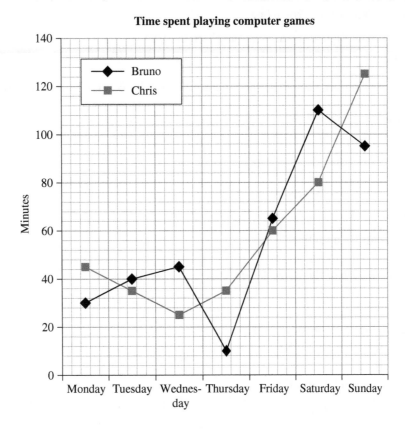

Time spent playing computer games

a How many minutes did Bruno spend playing computer games on Thursday?

_____ minutes

b Who spent more time on computer games on Saturday, and by how much?

_____ spent _____ more minutes on Saturday.

c Who spent more time on computer games during the whole week, and by how much?

_____ spent _____ more minutes during the week.

Stem-and-leaf diagrams

The stem goes to the left of the vertical line.

The leaves consist of one digit and you must put them in order.

1 The stem-and-leaf diagram shows the time in minutes that it takes for 15 students to get to school.

```
0 | 5 6 8 9
1 | 0 1 1 5 6 7 9
2 | 1 5 8
3 | 2
```

Key: 2 | 1 represents 21 minutes

a What was the length of the longest journey?

_____ minutes

b What was the range of journey times?

_____ minutes

c What was the modal journey time?

_____ minutes

d What was the median journey time?

_____ minutes

2 Below are the scores of 30 students in a quiz.

14	35	44	37	19	32	28	29	31	33
19	27	38	41	32	28	19	25	20	31
24	30	19	27	27	36	42	37	27	19

a Show this information in a stem-and-leaf diagram.

Key: ___ | ___ represents _____

b Find the median score.

c Find the range of scores.

d Find the modal score.

3 This stem-and-leaf diagram showing the masses of 20 parcels has four pieces of missing data.

```
1 | 6  6  8  9  9
2 | 0  1  2  2  __  __  7  9
3 | 4  6  __
4 | 2  3  5  __
```

Key: 2 | 1 represents 21 kg

a Use the information below to complete the table.

The mean is 28 kg.
The median is 23 kg.
The mode is 22 kg.
The range is 32 kg.

b Ten more parcels were added to the collection.
Their masses, in kilograms, were:

19   25   34   18   25   44   28   33   37   41

Show the masses of all 30 parcels in the stem-and-leaf diagram below.

```
1 |
2 |
3 |
4 |
```

Key: 2 | 1 represents 21 kg

Exercise 1 · Circles

Student book topic 11.1

Circumference of a circle $= \pi d = 2\pi r$

In all calculations, use the π key on your calculator.

Give all answers correct to 1 decimal place.

1 Put the correct labels on these parts of a circle.

Arc   Chord   Circumference   Diameter   Radius

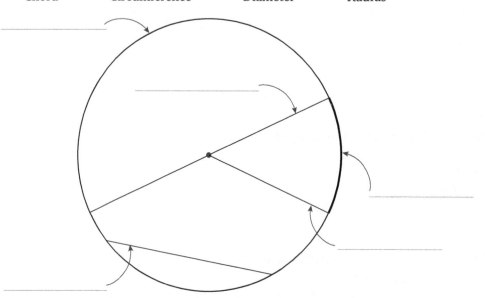

2 A circle has a radius of 6 cm.

Work out the circumference of the circle.

_____ cm

3 A circular table has a radius of 60 cm.
Find the circumference of the table.

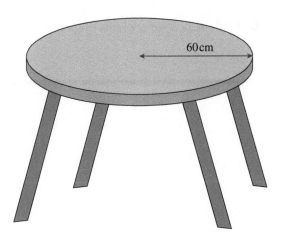

_____ cm

4 A piece of ribbon of length 65 cm just wraps around a can.
Work out the diameter of the can.

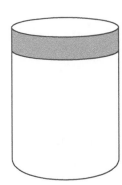

_____ cm

5 Find the perimeter of this shape.

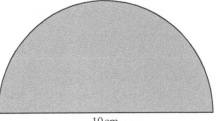

10 cm

_____ cm

6 Find the perimeter of this shape.

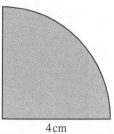

4 cm

_____ cm

7 Find the perimeters of these shapes.

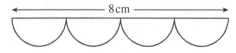

8 cm

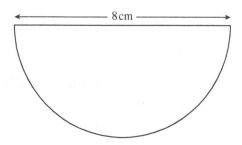

8 cm

Perimeter = _____ cm   Perimeter = _____ cm

Area of a parallelogram

Area of a parallelogram = base × perpendicular height

1 Work out the area of this parallelogram.

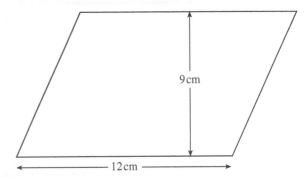

Area = _____ cm²

2 Complete the table below, which shows the length, perpendicular height and area of some parallelograms.

Length	Perpendicular height	Area
11 cm	7 cm	
8 cm		48 cm²
	9 cm	99 cm²
	6.4 cm	30.08 cm²

3 The two parallelograms below have the same area.
Find the value of x.

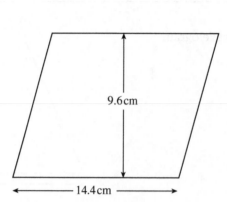 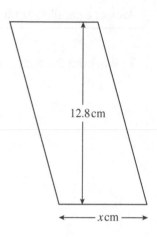

9.6 cm

14.4 cm

12.8 cm

x cm

$x =$ _____

4 By taking appropriate measurements,
find the area of this parallelogram.

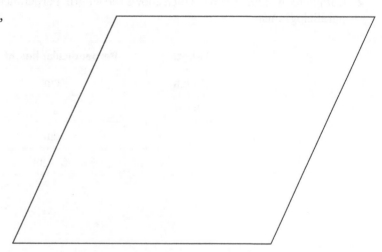

Area = _____ cm^2

5 Put a ring around the odd one out in the four parallelograms below.

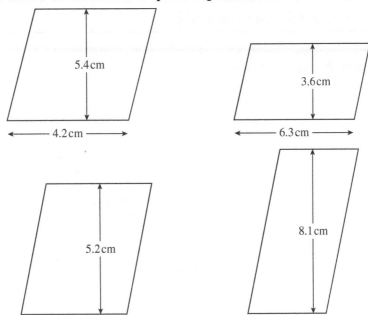

6 a A parallelogram has a perpendicular height equal to its base.
The area of the parallelogram is 134.56 cm².
Find the length of the base of the parallelogram.

_____ cm

b Another parallelogram has a base length equal to twice the perpendicular height.
The area of the parallelogram is 115.52 cm².
Find the length of the base of the parallelogram.

_____ cm

Area of a triangle

Area of a triangle $= \frac{1}{2} \times$ base $\times$ perpendicular height

1 Work out the areas of the triangles below.

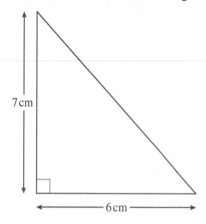

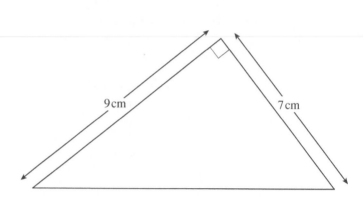

Area = _____ cm^2 Area = _____ cm^2

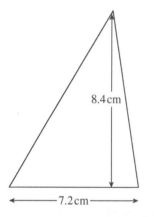

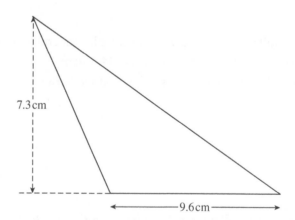

Area = _____ cm^2 Area = _____ cm^2

2 A triangle has a base of 7.6 cm and an area of 15.96 cm^2
Work out the perpendicular height of the triangle.

Perpendicular height = _____ cm

3 Take measurements of this triangle to work out the area.

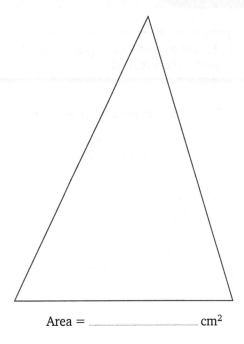

Area = _____ cm²

4 Draw lines to connect pairs of shapes with equal area.

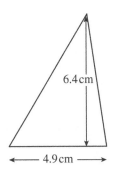 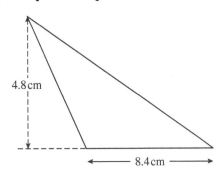 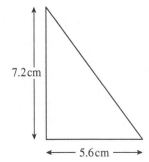

6.4 cm 4.9 cm

4.8 cm 8.4 cm

7.2 cm 5.6 cm

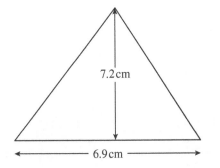 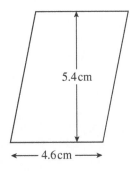 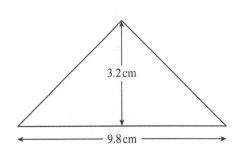

7.2 cm 6.9 cm

5.4 cm 4.6 cm

3.2 cm 9.8 cm

Area of a trapezium

> Area of a trapezium $= \frac{1}{2}(a + b)h$, where a and b are the lengths of the parallel sides and h is the perpendicular height.

1 Find the areas of the trapeziums below.

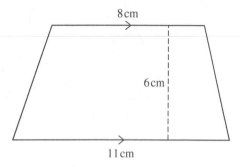

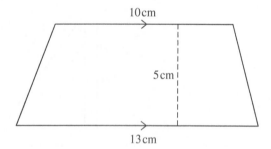

Area = _____ cm^2

Area = _____ cm^2

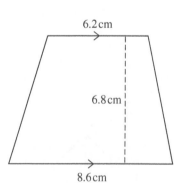

Area = _____ cm^2

2 A trapezium has parallel sides of length 7 cm and 8.4 cm.
The area of the trapezium is 52.36 cm².
Work out the perpendicular height of the trapezium.

_____ cm

3 This trapezium has an area of 47.04 cm².
Find the value of x.

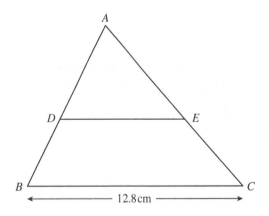

$x =$ _____

4 a The triangle *ABC* has an area of 61.44 cm².
Work out the perpendicular height of the triangle.

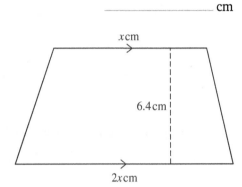

Perpendicular height = _____ cm

b The trapezium *BDEC* has a perpendicular height equal to that of triangle *ADE*.
BC is twice the length of *DE*.
Show that the area of the trapezium *BDEC* is three times the area of triangle *ADE*.

Area of a circle $= \pi r^2$

In all calculations, use the π key on your calculator.

Give all answers correct to 1 decimal place.

1 Work out the area of a circle with a radius of 8 cm.

Area = _____ cm^2

2 Work out the area of a circle with a diameter of 9.6 cm.

Area = _____ cm^2

3 A one euro coin has a diameter of 23.25 mm.
A two euro coin has a diameter of 25.75 mm.

  a Find the area of one face of a one euro coin.

Area = _____ mm^2

  b Find the area of one face of a two euro coin.

Area = _____ mm^2

  c By how much is the area of one face of the two euro coin bigger than the one euro coin?

It is _____ mm^2 bigger.

4 Take measurements to work out the areas of these shapes.

a

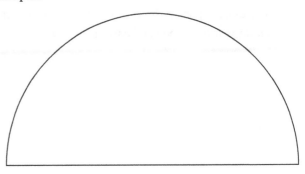

Area = _____ cm²

b

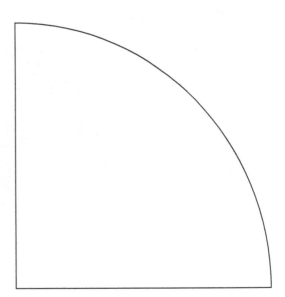

Area = _____ cm²

5 A CD has a diameter of 12 cm.
The unused part at the centre has a radius of 2 cm.
Find the area of the used part of the CD.

Area of used part is _____ cm²

6 This shape is made of three semicircles of diameter 10 cm and one of diameter 30 cm.
Work out the shaded area.

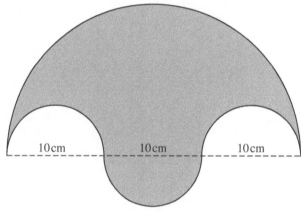

10 cm   10 cm   10 cm

Shaded area = _____ cm²

Compound areas are found by adding individual areas together.

The surface area of a solid is equal to the area of its net.

1 Find the areas of the shapes below.

a

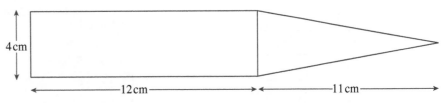

Area = _____ cm²

b

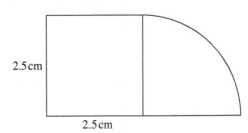

Area = _____ cm²

2 This shape is made of a square of side 6 cm and a trapezium of height 6 cm. The area is 63 cm².
Find the length of the top of the shape.

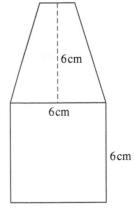

Length = _____ cm

3 Find the volume of the shapes below.

a

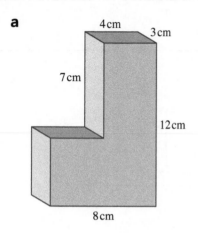

4 cm

3 cm

7 cm

12 cm

8 cm

b

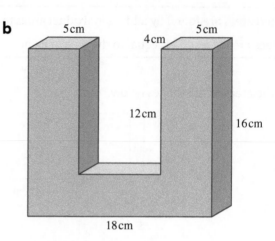

5 cm

4 cm

5 cm

12 cm

16 cm

18 cm

_____ cm³   _____ cm³

4 a What is the name of the 3-D shape that the net shown would make?

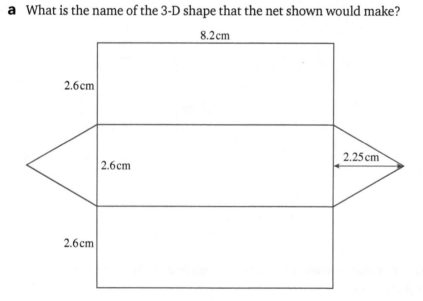

8.2 cm

2.6 cm

2.6 cm

2.25 cm

2.6 cm

_____

b Find the surface area of the solid.

Surface area = _____ cm²

5 a Draw an accurate net of this cuboid.
   It has been started for you.

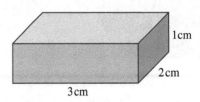

b Find the surface area of the net.

Surface area = _____ cm²

c Calculate the volume of the cuboid.

_____ cm³

6 Here is the net of a square-based pyramid.
Calculate the surface area of the net.

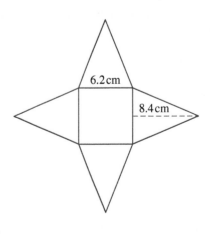

_____ cm²

Derive a formula means 'find and write down a formula'.

1 Match the costs, C, on the left with the correct formula on the right.

4 oranges at x cents each and 3 lemons at y cents each	$C = 3x + 4y$
4 adult tickets at $\$y$ and 3 child tickets at $\$x$	$C = 12 + xy$
4 cartons of milk, each containing x litres, at y cents per litre	$C = 4x + y$
4 books at $\$3$ each and x books at $\$y$ each	$C = 4x + 3y$
4 drinks at x cents and one at y cents	$C = 4xy$

2

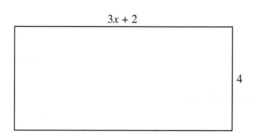

$3x + 2$

4

a Write down a formula, in its simplest form, for the perimeter, P, of the rectangle.

$P =$ _____

b Write down a formula for the area, A, of the rectangle.

$A =$ _____

3 Faith earns $m per month.
Henrik earns $n per month.
They rent an apartment that costs $a per month.
Write a formula for $R, the amount they have left each month after they have paid the rent.

$R =$  _____

4

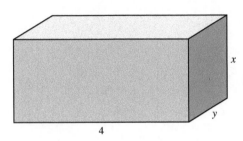

a Write a formula for the volume, V, of the cuboid.

$V =$ _____

b Write a formula for the surface area, S, of the cuboid.

$S =$ _____

5 A taxi company charges a fee of $4 plus $0.50 per kilometre.
Write a formula for $C, the cost of a journey of d km.

$C =$ _____

Replace the letters in a formula with given numbers.

Use BIDMAS.

Use the laws of negative numbers.

1 The perimeter P cm, of a rectangle is given by the formula $P = 2(l + w)$

Find the value of P when $l = 6$ and $w = 5$

$P =$ _____

2 If $S = 2a - 3b$, find the value of S when $a = 4$ and $b = 3$

$S =$ _____

3 The surface area, S cm², of a cube with sides of x cm is $S = 6x^2$

Find S when $x = 2$

$S =$ _____

4 The formula for the surface area, $A\,\text{cm}^2$, of a sphere, is given by the formula
$A = 4\pi r^2$, where r is the radius.
Find the value of A when $r = 6$

$A = $ _____

5 $v = u + at$
Find the value of v when $u = 6$, $a = -2$ and $t = 3$

$v = $ _____

6 If $f = \dfrac{ab}{a + b}$, find f when $a = 4$ and $b = 4$

$f = $ _____

7 If $t = \dfrac{v - u}{a}$, find t when $a = -4$, $v = 12$ and $u = 20$

$t = $ _____

8 The area, $A\,\text{cm}^2$, of a regular hexagon of side length s cm, is given
by the formula $A = 2.6s^2$
Find the area of a hexagon with sides of 5 cm.

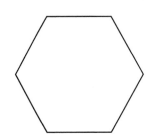

$A = $ _____

Sometimes after substitution you have to solve an equation.

1 Complete the working out to find the value of x when $A = 24$ and $y = 2$ in the formula $A = 3xy$

$A = 3xy$

_____ $= 3 \times x \times$ _____

_____ $=$ _____ x

$x =$ _____

2 $P = 3a + 2b$

Find the value of a when $P = 25$ and $b = 2$

$a =$ _____

3 $s = ut + \frac{1}{2}at^2$

Find the value of u when $s = 92$, $a = 10$ and $t = 4$

$u =$ _____

4 $p = 2s + \frac{1}{t}$

Find the value of s when $p = 1$ and $t = 3$

$s =$ _____

5 $x = 3(a - 2b)$

Find the value of a when $x = 36$ and $b = -2$

$a =$ _____

6 $I = \frac{PRT}{100}$

Find R if $I = 72$, $P = 400$ and $T = 2$

$R =$ _____

Transformations

Translations, rotations and reflections produce congruent shapes.

1

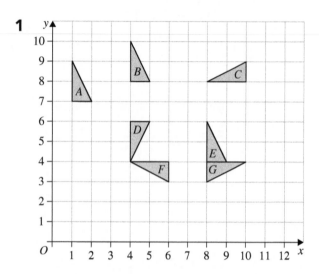

Match the descriptions with the transformations.

A to B
B to D
D to F
E to G
B to E
F to G
C to G

90° clockwise rotation about (4, 4)
180° rotation about (9, 6)
Translation 3 to the right and 1 up
Reflection in x = 7
Reflection in y = 7
Translation 4 to the right and 4 down
90° clockwise rotation about (8, 4)

2 a Reflect shape *A* in the line. Label the new shape *B*.

b Rotate shape *A* 90° anticlockwise about the point *X*. Label the new shape *C*.

c Shape *B* can be reflected on to shape *C*. Draw in the mirror line.

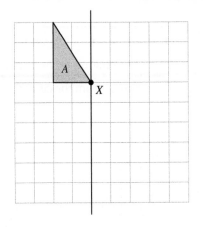

3 a Translate shape *A* 4 squares left and 2 squares down. Label the new shape *B*.

b Reflect shape *B* in the line $y = 4$. Label the new shape *C*.

c Rotate shape *C* 180° about the point (7, 3). Label the new shape *D*.

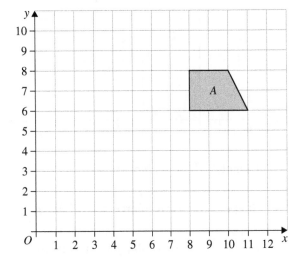

4 Mark reflects shape *A* in the line $x = 5$ and then rotates the image 90° anticlockwise about (5, 6).

Shirley rotates shape *A* 90° clockwise about (5, 6) and then reflects the image in the line $x = 5$

Use the grid to find out if their final images are in the same place.

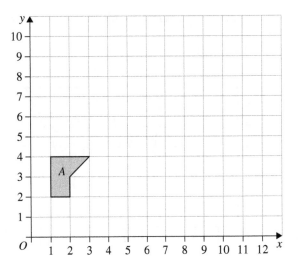

Enlargements produce similar shapes.

All measurements are multiplied by the same scale factor.

The distance from the centre of enlargement to each point is multiplied by the scale factor.

1

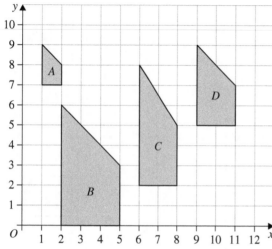

Complete these statements.

Shape _____ is an enlargement of shape _____ with a scale factor 3.

Shape _____ is an enlargement of shape _____ with a scale factor 2.

Shape _____ is not an enlargement of any of the other shapes.

2

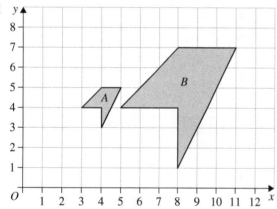

Complete the statement below.

B is an enlargement of *A* with a scale factor of _____, centre of enlargement (_____ , _____).

3 On the grid below plot and join the points (4, 1), (2, 3) and (5, 4). Label the triangle *A*.
Plot and join the points (2, 9), (0, 3) and (−4, 7). Label the triangle *B*.

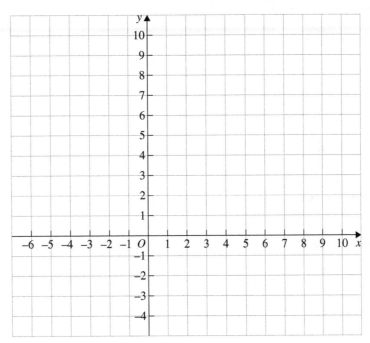

Complete the sentence:

Triangle *B* is an enlargement of triangle *A*, scale factor _____ and centre (_____ , _____).

4
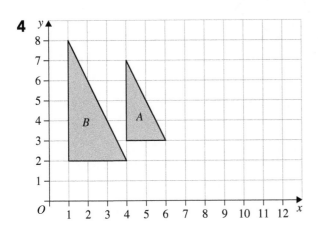

Complete the statement below.

B is an enlargement of *A* with a scale factor of _____, centre of enlargement (_____ , _____).

5

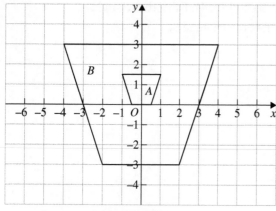

Complete the statement below.

B is an enlargement of *A* with a scale factor of _____, centre of enlargement (_____ , _____).

6 Draw an enlargement of this shape, scale factor 2, with *C* as the centre of enlargement.

C
●

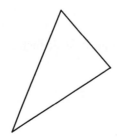

7 Draw an enlargement of this shape, scale factor 3, with *A* as the centre of enlargement.

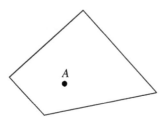

8

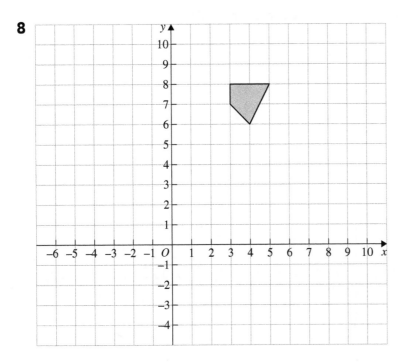

Enlarge the shape with a scale factor 3, centre of enlargement (7, 9).

9

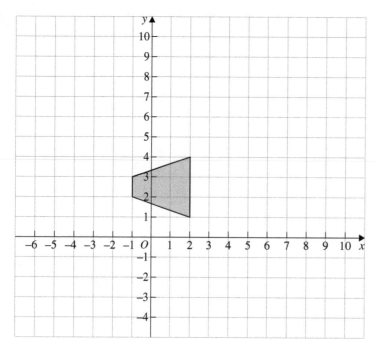

Enlarge the shape with a scale factor 4, centre of enlargement (0, 3).

 10

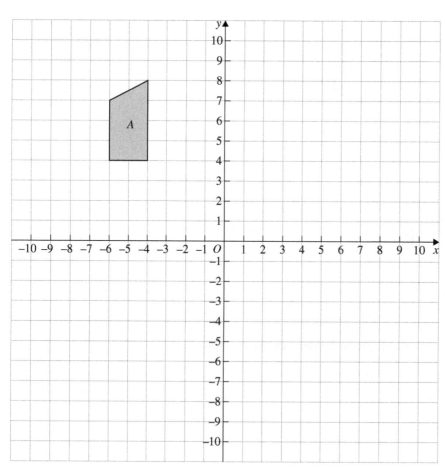

a Enlarge shape *A* by a scale factor of 3, centre (−8, 7).
Label the new shape *B*.

b Translate *B* 6 squares to the left and 6 squares down.
Label the new shape *C*.

c Find the centre of enlargement that transforms shape *A* directly to shape *C*.

Centre of enlargement = (_____ , _____)

d Find the area of shape *A*.

Area of shape *A* = _____ squares

e Find the area of shape *B*.

Area of shape *B* = _____ squares

Midpoints

The midpoint of a line joining (a, b) and (c, d) is at $\left(\dfrac{a+c}{2}, \dfrac{b+d}{2}\right)$

1 a Find the midpoint of the line AB by calculation.
Check your answer by measuring on the diagram.

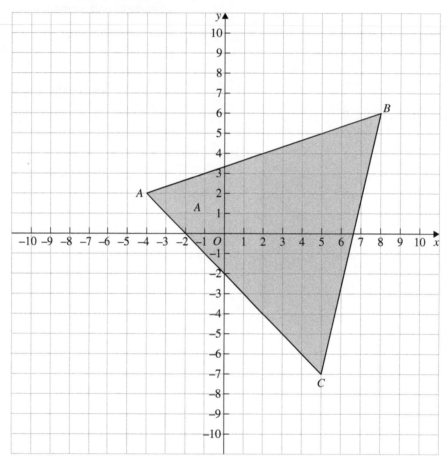

The midpoint of AB is (_____ , _____)

b Calculate the midpoint of BC.

The midpoint of BC is (_____ , _____)

2 Calculate the midpoint of the line joining $(5, 3)$ and $(-2, 8)$.

The midpoint is (_____ , _____)

3 A is the point $(3, -4)$.
M, the midpoint of AB, is at $(5, 1)$.
Find the coordinates of B.

B is the point (_____ , _____)

4 A parallelogram $ABCD$ joins $A(4, -3)$, $B(6, 1)$, $C(1, 11)$ and D.

a Calculate the coordinates of M, the midpoint of the diagonal AC.

M is the point (_____ , _____)

b M is also the midpoint of the diagonal BD.
Use this fact to find the coordinates of D.

D is the point (_____ , _____)

Linear sequences of numbers

A linear sequence has equal steps between the terms.

1 Write down the next three terms of these linear sequences.

3, 7, 11, 15, ———, ———, ———

2, 9, 16, 23, ———, ———, ———

15, 13, 11, 9, ———, ———, ———

14, 11, 8, 5, ———, ———, ———

2 Fill in the gaps in these linear sequences.

4, 7, ———, 13, ———, ———, 22

9, ———, 19, 24, ———, ———, 39

6, ———, ———, ———, ———, 41

———, 27, ———, ———, ———, −1

3 The first term in a sequence is 5.
The term-to-term rule is 'add 7'.

a Find the third term of the sequence.

b Find the tenth term of the sequence.

4 The first term in a sequence is 12.
The term-to-term rule is 'add 4'.

   a Find the third term of the sequence.

   b Find the hundredth term of the sequence.

5 Match the descriptions on the left with the position-to-term rules on the right.

2nd term is 12, 5th term is 21
3rd term is 12, 5th term is 20
1st term is 12, 5th term is 20
4th term is 12, 5th term is 20
4th term is 12, 5th term is 21

Term = 2 × position number plus 10
Term = 8 × position number minus 20
Term = 3 × position number plus 6
Term = 9 × position number minus 24
Term = 4 × position number

6 a Write down the first five terms of the sequence where the position-to-term rule is:
term = 4 × position number plus 2

   ————, ————, ————, ————, ————

   b Write down the first five terms of the sequence where the position-to-term rule is:
term = 2 × position number plus 3

   ————, ————, ————, ————, ————

   c Explain why there will never be any numbers that are in both sequences.

Linear sequences from patterns of shapes

Use a table to show the information from the patterns.

1 Look at these patterns of squares.

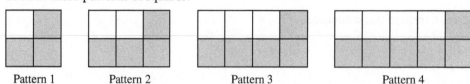

Pattern 1   Pattern 2   Pattern 3   Pattern 4

a Complete the table:

Pattern number	1	2	3	4	5	6
Number of white squares	1					
Number of grey squares	3					
Total number of squares	4					

b What is the number of white squares in pattern 50?

....

c What is the number of grey squares in pattern 50?

....

d What is the total number of squares in pattern 50?

....

e Complete the statements:

Number of grey squares = pattern number + _____

Total number of squares = _____ × pattern number + _____

2 Look at these patterns of triangles.

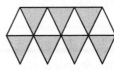

Pattern 1   Pattern 2   Pattern 3   Pattern 4

a Complete the table:

Pattern number	1	2	3	4	5	6
Number of white triangles	1					
Number of grey triangles	1					
Total number of triangles	2					

b What is the number of grey triangles in pattern 100?

_____

c What is the total number of triangles in pattern 100?

_____

d Complete the statement:

Total number of triangles = _____ × pattern number − _____

3 Look at these patterns of circles.

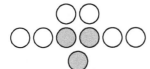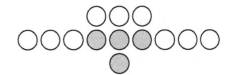

Pattern 1   Pattern 2   Pattern 3

a What is the number of grey circles in pattern n?

_____

b What is the number of white circles in pattern n?

_____

c What is the total number of circles in pattern n?

_____

The *n*th term is the rule for finding the number in position *n*.

It is a position-to-term rule, not a term-to-term rule.

1 Write down the first four terms of the sequence whose *n*th term is $3n - 1$.

————, ————, ————, ————

2 Write down the first four terms of the sequence whose *n*th term is $11 - 3n$.

————, ————, ————, ————

3 Look at this sequence:

Position	1	2	3	4	5
Sequence	5	9	13	17	21

Complete the following statements:

The term-to-term rule is + ————.

The sequence is related to the ———— times table.

The *n*th term is ———— n + ————.

The 100th term is ————.

4 Match the sequences with the *n*th terms.

2, 5, 8, 11, 14, …
5, 7, 9, 11, 13, …
−2, −1 , 0, 1, 2, …
5, 8, 11, 14, 17, …
4, 7, 10, 13, 16, …
2, 1, 0, −1, −2, …
4, 5, 6, 7, 8, …

$2n + 3$
$3 - n$
$3n - 1$
$n + 3$
$3n + 2$
$3n + 1$
$n - 3$

5 Find the *n*th term and the 50th term for each of these sequences.

  a −5, −1, 3, 7, 11, …

*n*th term = _____

50th term = _____

  b 7, 8.5, 10, 11.5, 13, …

*n*th term = _____

50th term = _____

  c 20, 14, 8, 2, −4, …

*n*th term = _____

50th term = _____

> Probability of a successful outcome = $\dfrac{\text{number of successful outcomes}}{\text{total number of outcomes}}$
>
> Probability an outcome does not happen = 1 − probability the outcome happens

1 Jo Clarke has some tiles that spell her name.

J O   C L A R K E

  a She puts them in a bag and chooses one at random.
     What is the probability that she chooses the letter A?

Her sister Rachel also has a set of tiles.

R A C H E L C L A R K E

  b She puts them in a bag and chooses one at random.
     What is the probability that Rachel chooses the letter A?

2 A bag contains 1 white, 3 black and 6 blue beads.
  What is the probability of a bead selected at random being:

  a white

  b blue

  c not white

  d black or white?

3 Mahinda is playing a game. He is trying to throw a ball into a jam jar.
There are two different sizes of jam jar.

The probability that the ball lands in a large jam jar is $\frac{2}{5}$

The probability that the ball lands in a small jam jar is $\frac{1}{10}$

a Find the probability that the ball does not land in a large jam jar.

_____

b Find the probability that the ball does not land in a small jam jar.

_____

c Mahinda says that the probability that the ball does not land in a jam jar at all is $\frac{1}{2}$
Explain why Mahinda is correct.

_____

4 A sports club has 50 members.
Some are adult members (at least 18 years old), others are youths (under 18 years old).
The table shows the numbers of each type of member.

	Male	Female
Youth	11	16
Adult	13	10

A member is chosen at random.
What is the probability that the chosen member is:

a male

_____

b adult

_____

c female youth?

_____

Listing outcomes

You should list outcomes in a systematic way to avoid errors and omissions.

1 Jodi flips two coins.

a List all the possibilities in the table.

First coin	Second coin
Head	Head

Write down the probability that:

b both coins land on heads

_____

c at least one coin lands on heads.

_____

2 A chef makes pies using four different types of filling: mushroom, lamb, vegetable and chicken.
He makes each pie with two types of pastry: puff pastry and shortcrust pastry.

a He makes a list of all the different types of pie that he makes.
Complete the list.

Filling	Pastry
Mushroom	Puff

A customer chooses a pie at random from the list.
What is the probability that the customer chooses:

b a vegetable pie

_____

c a pie with short-crust pastry

_____

d a chicken pie with puff pastry?

_____

3 Beki spins these spinners.

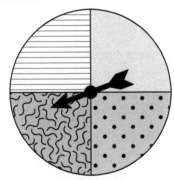 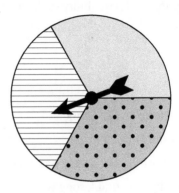

a List all the possible combinations.

Left spinner	Right spinner
Stripes	Stripes

b What is the probability that both spinners land on the same pattern?

> Experimental probability = $\dfrac{\text{number of successful trials}}{\text{total number of trials}}$

1 Tina has a bag containing red balls, green balls and yellow balls.

She chooses a ball at random, notes the colour and puts it back.

She does this 60 times.

Here are her results.

Red	Green	Yellow
26	14	20

What is the experimental probability that she chooses a green ball?

The experimental probability is _____ .

2 Samira listens to music on her phone.

The songs are chosen at random.

There is a probability of $\frac{1}{3}$ that a song has a female singer.

There is a probability of $\frac{3}{5}$ that a song will be longer than 3 minutes.

Samira is going to listen to 60 songs.

a How many songs by female singers does she expect to hear?

b How many songs longer than 3 minutes does she expect to hear?

3 Holly spins this spinner 40 times. Here are her results.

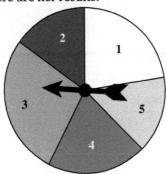

Number	1	2	3	4	5
Frequency	9	6	11	8	6

a Complete this table showing her experimental probabilities.

Number	1	2	3	4	5
Experimental Probability	$\frac{9}{40}$			$\frac{1}{5}$	

b Jaya spins the same spinner 100 times. Here are her results.

Number	1	2	3	4	5
Frequency	21	17	25	21	16

What is Jaya's experimental probability of scoring a 5?

_____

c Whose probabilities do you think are more accurate?
Give a reason for your answer.

_____ 's probabilities are more accurate because

_____

4 A factory makes circuit boards for phones.
The manager randomly tests 40 circuit boards.
Three are faulty.
The factory makes 2000 circuit boards in a day.
How many are likely to be faulty?

_____

_____

A function maps one set of numbers onto another.

An inverse function reverses the process.

1 Complete these mapping diagrams.

a
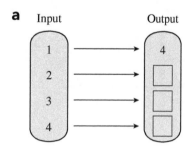
Rule: $x \rightarrow x + 3$

b Input   Output
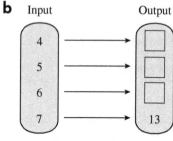
Rule: $x \rightarrow 2x - 1$

c
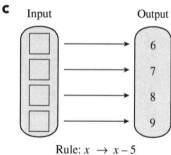
Rule: $x \rightarrow x - 5$

d
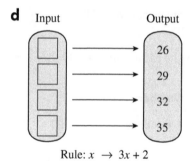
Rule: $x \rightarrow 3x + 2$

2 Find the rule for these mapping diagrams.

a
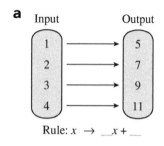
Rule: $x \rightarrow \_\_x + \_\_$

b Input   Output
1 → 3
2 → 8
3 → 13
4 → 18
Rule: $x \rightarrow \_\_x - \_\_$

3 Complete these mapping diagrams.

a

Input   Output

3
5
8
11

Rule: $x \rightarrow 3x - 7$

b

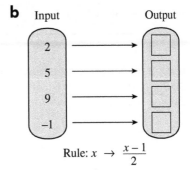

Input   Output

2
5
9
−1

Rule: $x \rightarrow \dfrac{x-1}{2}$

c

Input   Output

1
4
9
−2

Rule: $x \rightarrow \dfrac{x}{2} - 5$

d

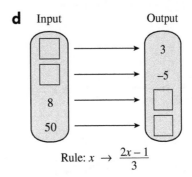

Input   Output

3
−5

8
50

Rule: $x \rightarrow \dfrac{2x-1}{3}$

4 Find the rule for these mapping diagrams.

a

Input   Output

1   6
3   10
5   14
7   18

Rule: $x \rightarrow \underline{}x + \underline{}$

b

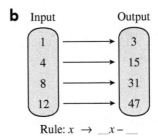

Input   Output

1   3
4   15
8   31
12   47

Rule: $x \rightarrow \underline{}x - \underline{}$

Functions

The function machine $x \longrightarrow \boxed{\times 3} \longrightarrow \boxed{-1} \longrightarrow y$ can be written as $y = 3x - 1$

1 Write these function machines as equations in the form $y = \dots$

a $x \longrightarrow \boxed{\times 2} \longrightarrow \boxed{+4} \longrightarrow y$

$y = \underline{\hspace{4cm}}$

b $x \longrightarrow \boxed{\times 4} \longrightarrow \boxed{-5} \longrightarrow y$

$y = \underline{\hspace{4cm}}$

c $x \longrightarrow \boxed{+3} \longrightarrow \boxed{\times 2} \longrightarrow y$

$y = \underline{\hspace{4cm}}$

d $x \longrightarrow \boxed{-5} \longrightarrow \boxed{\div 2} \longrightarrow y$

$y = \underline{\hspace{4cm}}$

e $x \longrightarrow \boxed{\times 3} \longrightarrow \boxed{-5} \longrightarrow \boxed{\div 2} \longrightarrow y$

$y = \underline{\hspace{4cm}}$

2 Complete the function machine for each of these equations.

a $y = 2x + 7$

b $y = \dfrac{x - 3}{4}$

c $y = \dfrac{3x - 5}{4}$

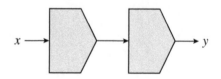

3 Complete the function machine and table of values for each of these equations.

a $y = 4x - 3$

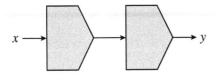

x	2		7	
y		13		37

b $y = \dfrac{x + 2}{3}$

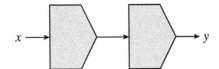

x	4		13	
y		4		8

c $y = \dfrac{2x + 2}{3}$

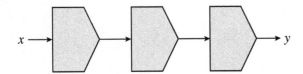

x	8		−7	
y		4		−8

4 Work out the functions for these tables of x and y values.

a

x	1	2	3	4
y	4	6	8	10

$y =$ _____

b

x	1	2	3	4
y	6	14	22	30

$y =$ _____

Linear graphs

Complete a table of values to help draw a linear graph.

1 Complete the table and use it to draw the graph of $y = x + 2$

x	−3	1	3
y	−1		

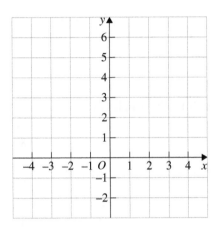

2 Complete the table and use it to draw the graph of $y = 2x − 1$

x	−3	1	3
y			

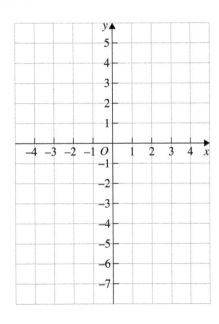

3 Complete the table and use it to draw the graph of $y = 2 - x$

x	-3	0	4
y			

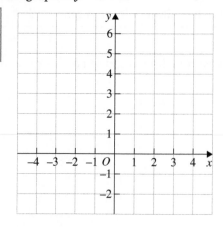

4 Complete the table and use it to draw the graph of $y = \frac{1}{2}x + 3$

x	-6	0	6
y			

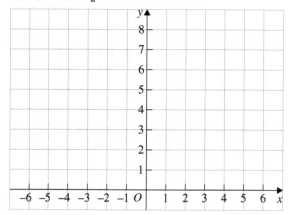

5 Complete the grid to match the points and the graphs of the equations.
Each point lies on at least two graphs.

Equation	(6, 13)	(1.5, 5.5)	(−4, 0)	(3, 7)	(5, 10)	(2, 4)	(3, 4)
$y = x + 4$	✗	✔					
$y = 2x$							
$y = 2x + 1$							
$y = \frac{1}{2}x + 2$							
$y = 3x - 2$							
$y = 3x - 5$							
$y = 7 - x$							

To find the equation from a graph, first complete a table and then find the rule.

1 Here is a table for a graph.

x	2	3	4	5
y	8	13	18	23

a Write down the term-to-term rule for the values of y.

The rule is + _____

b Insert the multiplication line in the middle row of the table, and fill in the gaps by the arrows.

x	2	3	4	5
× ____				
y	8	13	18	23

× _____

− _____

c The equation of the line is $y =$ _____ $x -$ _____

2 Find the equation for this table.

x	1	2	3	4
× ____				
y	6	10	14	18

× _____

+ _____

The equation of the line is $y =$ _____ $x +$ _____

3 Find the equation for this table.

x	−2	−1	0	1
× ____				
y	−4	−1	2	5

× _____

+ _____

The equation of the line is $y =$ _____ $x +$ _____

4 Complete the table to find the equation of this graph.

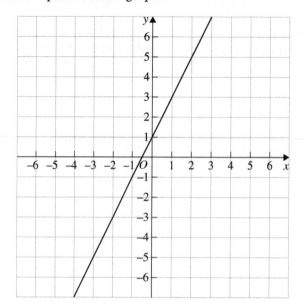

x	0	1	2	3
$\times$ _____				
y	1			

$\times$ _____

$+$ _____

The equation is $y =$ _____

5 Complete the table to find the equation of this graph.

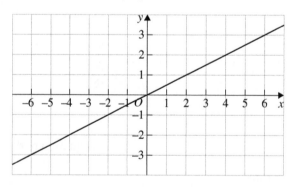

x	0	1	2	3
y				

The equation is $y =$ _____

6 Complete the table to find the equation of this graph.

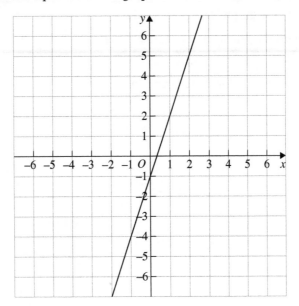

x				
$\times$ ___				
y				

The equation is $y =$ _____

A positive value of m means the graph has a positive slope.

A negative value of m means the graph has a negative slope.

1 Write down the values of m and c for these equations.

a $y = 2x + 5$

$m =$ _____, $c =$ _____

b $y = 3x - 2$

$m =$ _____, $c =$ _____

c $y = \frac{1}{2}x + 5$

$m =$ _____, $c =$ _____

d $y = 5x$

$m =$ _____, $c =$ _____

e $y = 7 - 3x$

$m =$ _____, $c =$ _____

2 Put a ring around the equations whose lines have a positive slope.
Underline the equations whose lines have a negative slope.

$y = 3x + 5$ $y = -x + 1$ $y = 6 + 2x$

$y = 4 - x$ $y = x - 5$ $y = -4$

$y = 4$ $y = x$ $y = 6 - 2x$

3

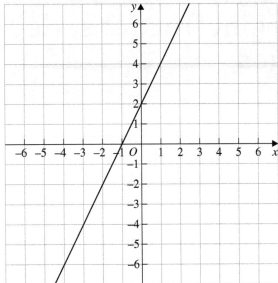

a Use the table to find the equation of this line.

x	−2	−1	0	1
× _____				
y	−2			

The equation is $y =$ _____

b Draw the reflection of the line in the y-axis.

c Use the table to find the equation of the line in part **b**.

x	0	1	2	3
× _____				
y	2			

The equation is $y =$ _____

d What can you say about the value of c in both equations?

_____

e What can you say about the value of m in both equations?

_____

_____

Finding equivalent fractions, decimals and percentages

To change a fraction to a decimal, divide the numerator by the denominator.

To change a percentage to a decimal, divide by 100.

To change a fraction or decimal to a percentage, multiply by 100.

To change a percentage to a fraction, divide by 100.

1 Match the fractions, decimals and percentages.

Fractions	Decimals	Percentages
$\frac{1}{2}$	0.75	65%
$\frac{13}{20}$	0.375	70%
$\frac{3}{4}$	0.48	50%
$\frac{9}{16}$	0.5	56.25%
$\frac{3}{5}$	0.7	48%
$\frac{12}{25}$	0.65	75%
$\frac{3}{8}$	0.07	37.5%
$\frac{7}{10}$	0.5625	60%
$\frac{7}{100}$	0.55	7%
$\frac{11}{20}$	0.6	55%

2 Change $\frac{4}{5}$ to a decimal and a percentage.

Decimal: _____   Percentage: _____

3 Change 0.64 to a percentage and a fraction in its lowest terms.

Percentage: _____   Fraction: _____

4 Change 87.5% to a decimal and a fraction in its lowest terms.

Decimal: _____   Fraction: _____

5 Write these in order of size, starting with the smallest.

a 0.45 $\frac{12}{25}$ 0.5   47%

_____ , _____ , _____ , _____

b 0.9   86%   0.845 $\frac{17}{20}$

_____ , _____ , _____ , _____

Increasing and decreasing by a percentage

To find a percentage of a quantity, change the percentage to a fraction and multiply.

1 Work these out.

  a 10% of $70

$ _____

  b 15% of $74

$ _____

  c 5% of 220 metres

_____ metres

  d 22.5% of $84

$ _____

2 There are 440 seats on a plane.
85% of the seats are occupied.
How many seats are occupied?

_____

3 An exam has 120 questions.
You need at least 70% to pass.
How many questions must you get correct?

_____

4 Increase $60 by 20%

$ _____

5 Decrease $140 by 35%

$ _____

6 Yasmin earns $40 000 per year.
She receives a 5% increase.
What is her new salary?

$ _____

7 A furniture shop advertises a table for $120.
The shop has a sale, with everything reduced by 30%.
The following week, the shop advertises

**AN EXTRA 20%
OFF SALE
PRICES!**

Work out the price of the table now.

$ _____

To find one quantity as a percentage of another, write it as a fraction first.

Percentage changes are always given as a percentage of the original amount.

1 In a school, there are 176 boys and 144 girls.

  a Find the total number of children in the school.

  _____

  b What percentage of the children are girls?

  _____ %

2 A ring weighs 8.5 grams.
3.4 grams of it is gold.
What percentage of the ring is gold?

  _____ %

3 The tables show the amount of fat and protein in fish and chicken breast.

Fish	
Fat	1 g
Protein	21 g
Other substances	94 g
Total	116 g

Chicken breast	
Fat	3 g
Protein	28 g
Other substances	55 g
Total	86 g

  a Work out the percentage of the fish that is protein.

  _____ %

  b Work out the percentage of the chicken that is fat.

  _____ %

4 Daniel and Leah buy a house for $160 000.
Three years later they sell it for $152 000.
Work out the percentage decrease in price.

_____%

5 Jodi rents a flat.
Her rent is $450 per month.
Carlos rents a flat.
His rent is $340 per month.
The landlord increases the rent.
Jodi now has to pay $475 per month.
Carlos now has to pay $360.

   a Work out Jodi's percentage increase.

_____%

   b Work out Carlos's percentage increase.

_____%

   c Who has had the larger actual increase in rent?

_____ has had the larger actual increase.

   d Who has had the larger percentage increase in rent?

_____ has had the larger percentage increase.

Identifying and collecting data

Student book topic 18.1

Chapter 6 covers processing data
Chapter 10 covers presenting data

> Sample sizes need to be sufficiently large to be accurate, but small enough to be manageable.
>
> Data needs to be collected in a way that avoids bias.

1 Explain why each of these samples would not give fair results.
In each case suggest a way of improving the survey.

a You do a survey of what students in your school think of the school meals.
You select a group of 20 students who eat a school meal every day.

Why the sample results will not be fair: _____

_____

_____

Improvement: _____

_____

_____

b A store manager wants to know what proportion of customers pay by cash rather than by credit card.
She collects sales receipts from the first 50 customers who make purchases on Monday morning.

Why the sample results will not be fair: _____

_____

_____

Improvement: _____

_____

_____

2 A newspaper headline reads

LOCAL LIBRARY HAS IMPROVED DURING THE LAST YEAR

a Here are some statistics from the library for last year and this year.
Put a tick next to those that you think show an improvement this year.

	Last year	This year	✓ or ✗
Number of fiction books	8675	8943	
Number of non-fiction books	1675	1599	
Number of new books added	602	586	
Number of music CDs	754	490	
Number of DVDs	698	872	
Number of newspapers/magazines	38	32	
Number of computers	12	21	
Number of people using the library	231	256	

b Why do you think the library is stocking fewer CDs and magazines?

_____

c Do you think the library has improved overall? Give a reason for your answer.

_____

Discrete data can only take certain values.

Continuous data can take any values.

1 Say whether each of these is an example of discrete or continuous data.

  a The number of parcels in a truck.

   This is _____ data.

  b The mass of all the parcels.

   This is _____ data.

  c The time it takes for the truck to deliver them to a sorting office.

   This is _____ data.

  d The volume of the parcels.

   This is _____ data.

2 This frequency table shows the time it took some students to complete a maths test.

Time, t minutes	Tally	Frequency
$5 < t \leqslant 10$	||	2
$10 < t \leqslant 15$	ᴎᴎ ᴎᴎ |||	13
$15 < t \leqslant 20$		17
$20 < t \leqslant 25$	ᴎᴎ |	
$25 < t \leqslant 30$	|||	3

  a Complete the table.

  b How many students took the test?

   _____ students

3 Russell measures the heights of his tomato plants.
   Here are his results.

43 cm	72 cm	56 cm	29 cm	38 cm	81 cm
56 cm	39 cm	60 cm	49 cm	77 cm	82 cm
36 cm	64 cm	70 cm	66 cm	59 cm	63 cm
55 cm	48 cm	75 cm	32 cm	38 cm	69 cm

Russell shows these data in a frequency chart. He starts like this:

Height (cm)	Tally	Frequency
20–30		
30–40		
40–50		
50–60		
60–70		
70–80		

He has made two different mistakes with the height groups.

Use the table below to complete an accurate frequency chart for the data.

Height, h cm	Tally	Frequency

4 The speeds in km/h of 50 cars were measured as follows:

62	54	56	73	78	63	68	70	66	60
54	58	65	55	57	69	67	61	64	63
53	56	68	76	57	48	57	68	82	68
78	72	75	65	67	64	54	58	62	74
67	80	87	69	74	78	70	76	46	67

Complete the frequency table to show these data.

Speed, s km/h	Tally	Frequency
$45 < s \leqslant 50$		
$50 < s \leqslant 55$		
$55 < s \leqslant 60$		
$60 < s \leqslant 65$		
$65 < s \leqslant 70$		
$70 < s \leqslant 75$		
$75 < s \leqslant 80$		
$80 < s \leqslant 85$		
$85 < s \leqslant 90$		

Two-way tables show data divided into two different sets of categories.

One set is shown horizontally and the other is shown vertically.

1 The table below shows information about the gender and age of the teachers in a school.

	Male	Female
22–35	6	14
36–45	13	7
Over 45	9	13

a How many teachers are there in the school?

b Which is the most common age group?

2 Here is a table showing information about people on a bus.

	0–3 years old	4–10 years old	11–16 years old	17–25 years old	26–50 years old	51 years or older
Male	2	3	9	2	1	9
Female	1	5	5	1	4	12

a How many of the people on the bus are less than 17 years old?

b How many of the females are 26 years old or older?

3 Olivier has some T-shirts.

The table shows the details.

	White	Red	Blue	Black
Long-sleeved	2	1	1	0
Short-sleeved	2	2	1	3

a How many T-shirts does Olivier have?

_____

b Olivier chooses a T-shirt at random.

What is the most likely colour?

_____

4 Complete this two-way table showing the number of left-handed and right-handed children in a class.

	Male	Female	Total
Left-handed	3		7
Right-handed		12	
Total			28

5 Rafad has 30 books.

They are either paperback or hardback.

They are either fiction or non-fiction.

He has 11 paperback fiction books.

He has 13 hardback books.

He has 14 non-fiction books.

Use this information to complete the table.

	Paperback	Hardback	Total
Fiction			
Non-fiction			
Total			

More ratio
Student book topic 19.1

Ratios can only compare quantities that are written in the same units.

Do not include units in a ratio.

Simplify ratios by dividing all parts by the same factor.

1 Complete the sentences below.

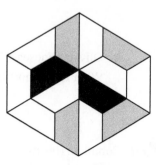

    a The ratio of black, grey and white trapeziums is 2 : _____ : _____

    b In its simplest form, the ratio is 1 : _____ : _____

2 Write the ratio of black, grey and white rectangles.
Give your answer in its simplest form.

_____ : _____ : _____

3 a How many cm are there in 1.5 m?

_____ cm

    b A rectangular frame is 1.5 m long and 80 cm wide.
Write the ratio of the length to the width.
Give your answer in its simplest form.

_____ : _____

4 Simplify the ratio $24:18:21:15$

_____ : _____ : _____ : _____

5 A bag contains 30 lemon sweets, 42 orange sweets and 24 banana sweets.
Write the ratio of lemon sweets to orange sweets to banana sweets.
Give your answer in its simplest form.

_____ : _____ : _____

6 Mia, Daniel, Camila and Mateo share some money in the ratio $2:3:1:4$

a What fraction of the money does Mia get?

_____

b What can you say about the amount of money Mia gets and the amount of money that Mateo gets?

_____

c What is the ratio of Daniel's share to Mateo's share?

_____ : _____

7 A recipe includes 2 teaspoons of clear honey and 1 tablespoon of lemon juice.
A teaspoon contains 5 ml and a tablespoon contains 15 ml.
Write the ratio of honey to lemon juice in its simplest form.

_____ : _____

8 A farmer has a rectangular field, 1.2 km long and 80 m wide.
He covers it with 5 cm of topsoil.
Write the ratio of the length to the width to the depth in its simplest form.

_____ : _____ : _____

Dividing in a given ratio

Ratio is about portions.

Find the size of each portion by dividing by the total of the ratios.

1 Divide 180 kg in the ratio 1:5:6

_____ kg, _____ kg, _____ kg

2 Divide $117 in the ratio 2:3:8

$ _____ , $ _____ , $ _____

3 Divide 184 minutes in the ratio 2:3:3

_____ minutes, _____ minutes, _____ minutes

4 $143 is divided in the ratio 2:4:5

Work out the difference between the largest share and the smallest share.

The difference is $ _____ .

5 Lola and Sanjay share $1000 between them in the ratio $4:1$
Lola shares her part between herself, her mother and her daughter in the ratio $2:1:1$
How much does her daughter receive?

Her daughter receives $ _____ .

6 The ratio of the ages of Gita, Anna and Shivi is $9:7:8$
Gita is 12 years older than Anna.
Work out Shivi's age.

Shivi is _____ years old.

7 In a bag of discs, the ratio of red discs to blue discs is $3:2$
The ratio of red discs to yellow discs is $2:1$
What is the ratio of blue discs to yellow discs?

_____ : _____

8 The angles of a triangle are in the ratio $4:7:9$
The largest angle is $45°$ bigger than the smallest angle.
Find the sizes of the three angles.

_____ °   _____ °   _____ °

To use the unitary method, you find the value of one item before finding the answer.

1 Marco buys 6 metres of pipe for $14.40.

a How much does 1 metre of pipe cost?

$ _____

b How much does 11 metres of pipe cost?

$ _____

2 5 litres of paint costs $7.60.
Work out the cost of 9 litres of paint.

$ _____

3 Lee, Jane and Wei share some money in the ratio 3 : 4 : 5
Wei receives $42 more than Lee.

a How much does Jane receive?

$ _____

b How much do they share?

$ _____

4 Six tickets to the theatre cost $102.
How much would 10 tickets cost?

$ _____

5 Here is a recipe for honeyed lamb stew for 4 people.
Next to it, write the recipe for 10 people.

Serves: 4	Serves: 10
2 tablespoons vegetable oil	_____ tablespoons vegetable oil
1200 g lamb stewing steak	_____ g lamb stewing steak
1 large onion, chopped	_____ large onions, chopped
1 green pepper, chopped	_____ green peppers, chopped
3 sticks celery, chopped	_____ sticks celery, chopped
4 large carrots, peeled and sliced	_____ large carrots, peeled and sliced
30 ml honey	_____ ml honey
25 ml fresh lemon juice	_____ ml fresh lemon juice
10 ml mustard	_____ ml mustard
50 ml tomato puree	_____ ml tomato puree
500 ml water	_____ ml water

6 Three tins of paint cover 2 walls with 2 coats.
How many tins are needed to cover 10 walls with 3 coats?

_____ tins

7 Hassan is 7 years old and Hope is 5 years old.
They share some money in the ratio of their ages.
Hassan receives $28 more than Hope.
If they share the same amount of money next year in the ratio of their ages, how much would they receive each?

Hassan will receive $ _____, Hope will receive $ _____.

Travel graphs

Student book topic 20.1

Graphs are a useful way of comparing two sets of data.

1 Asher drives 80 km from his home to his uncle's house.
Asher's son, Declan, makes the same journey by train.
Their journeys are shown on the graph below.

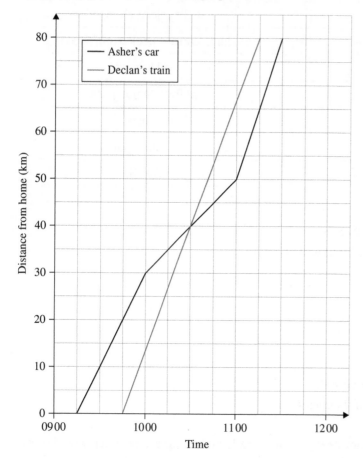

a At what time did Asher leave home?

b At what time did Declan's train leave?

c At what time were the car and the train the same distance from home?

d How far did the train travel in the first hour?

e At what time did Asher arrive at his uncle's house?

2 Sophia and Layla need to have a meeting.
The graph shows their journeys from their homes to the meeting.

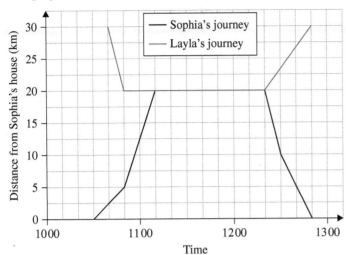

a How far apart do Sophia and Layla live?

_____ km

b Layla left later than Sophia. How much later?

_____ minutes

c How far from Layla's house did they meet?

_____ km

d Who arrived first, and by how much?

_____ arrived _____ minutes before _____.

e How long did the meeting last?

_____ minutes

f Whose speed was greater on the journey home? Explain how you know.

_____'s journey home was faster because _____

_____

3 Here is the timetable for the 07 13 train from Nakheel Harbour & Tower to Noor Islamic Bank.

07 13 Nakheel Harbour & Tower

07 17 Dubai Marina

07 22 Dubai Internet City

07 26 Mall of the Emirates

07 40 Noor Islamic Bank

The journey is shown on the graph below.

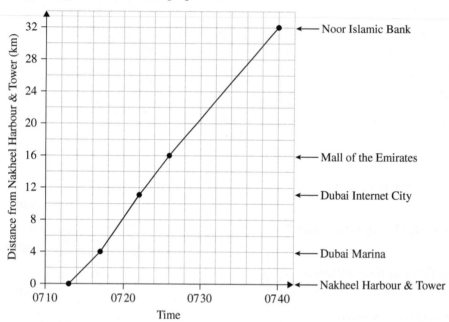

Here is the timetable for a train from Noor Islamic Bank to Nakheel Harbour & Tower.

07 16 Noor Islamic Bank

07 26 Mall of the Emirates

07 30 Dubai Internet City

07 35 Dubai Marina

07 40 Nakheel Harbour & Tower

a Add this journey to the graph.

b How far is it from Dubai Marina to Mall of the Emirates?

_____ km

c How long does it take the train from Noor Islamic Bank to reach Nakheel Harbour & Tower?

_____ minutes

d How far from Noor Islamic Bank are the trains when they pass?

_____ km

Graphs showing change always have time on the horizontal axis.

1 Aiden was ill at the beginning of March.
The graph shows his daily temperature.
His temperature returned to normal on 12th March.

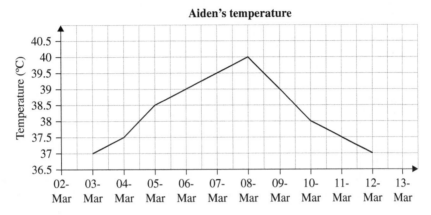

Aiden's temperature

a On what date was Aiden's temperature highest, and what was this temperature?

Aiden's temperature was _____ °C on _____ .

b For how many days was his temperature 37.5 °C or higher?

_____ days

c What is Aiden's normal temperature?

_____ °C

d How much did Aiden's temperature rise above the normal temperature?

_____ °C

2 The graph shows the amount (in thousands of litres) of two types of fuel sold each month at a local filling station.

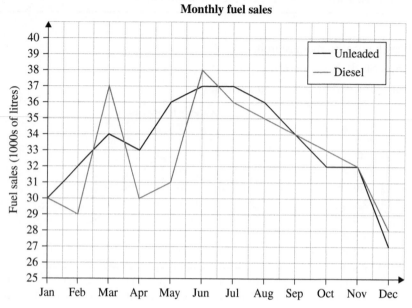

Monthly fuel sales

a What were the total sales of fuel in March?

_____ litres

b In which months were the sales of diesel greater than the sales of unleaded?

_____

c Calculate the total sales of unleaded petrol for the whole year.

_____ litres

d What percentage of the year's unleaded sales took place in May?

_____ %

3 Juan and Rosa are twins.

Their height was measured every year from the age of 2 to the age of 18.

Here is a graph showing Juan's height.

The table shows Rosa's height each year.

Age (years)	2	3	4	5	6	7	8	9	10
Height (cm)	89	94	101	106	114	120	127	134	140
Age (years)	11	12	13	14	15	16	17	18	
Height (cm)	148	153	158	162	163	163.5	164	164	

a Add Rosa's information to the graph.

b At what ages was Rosa taller than Juan?

_____

c Juan was 90 cm tall at the age of two.

How many years did it take Juan's height to increase by 50%?

_____ years

Set notation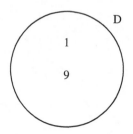

> ∈ means 'is an element of', ∉ means 'is not an element of', ∅ means the empty set,
>
> ∩ means the intersection of sets, ∪ means the union of sets and ⊂ means 'is a subset of'.
>
> Venn diagrams help display information about sets.

1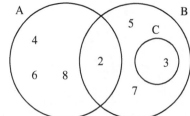

a Match the descriptions on the left with the sets on the right.

A ∩ B
B ∪ D
A ∪ B
D ∩ B
A ∪ C
C ∩ B

{2, 3, 4, 5, 6, 7, 8}
{2, 3, 4, 6, 8}
{2}
{3}
{1, 2, 3, 5, 7, 9}
{ }

b For each statement, say whether it is true or false.

C ⊂ B _____

C ∪ D = ∅ _____

B ∩ C = C _____

4 ∈ A ∪ D _____

A = {even numbers < 10} _____

B ∪ D = {odd numbers < 10} _____

B = {prime numbers < 10} _____

D = {square numbers < 10} _____

2

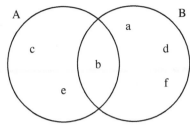

List the members of:

a A = { _____ }

b B = { _____ }

c A ∩ B = { _____ }

d A ∪ B = { _____ }

3 A = {factors of 12}
B = {multiples of 3 < 15}

a List the members of:

A = { _____ }   B = { _____ }

b Complete the Venn diagram:

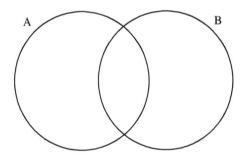

4 Complete the statements about this diagram:

a B ⊂ _____

b B ∩ _____ = ∅

c B ∪ C _____ A

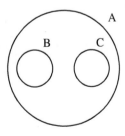

ξ is the universal set, containing every element in a particular problem.

The complement of set A, written as A', is the set of all elements not in A.

1 List the elements of the sets listed below.

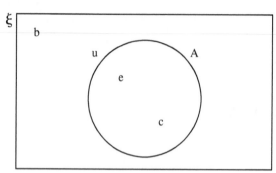

ξ = { _____ }

A = { _____ }

A' = { _____ }

2 ξ

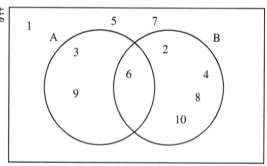

Match the descriptions to the sets.

A'
ξ
B'
(A ∪ B)'
(A ∩ B)'
A' ∩ B
A ∪ B'
ξ'

{1, 2, 3, 4, 5, 6, 7, 8, 9, 10}
{1, 2, 3, 4, 5, 7, 8, 9, 10}
{1, 2, 4, 5, 7, 8, 10}
{1, 3, 5, 6, 7, 9 }
{1, 3, 5, 7, 9}
{ }
{2, 4, 8, 10}
{1, 5, 7}

3 The Venn diagram shows the sports played by some children.

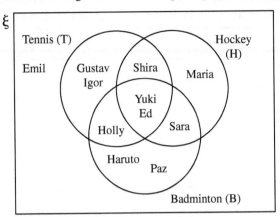

a Who does not play tennis, hockey or badminton?

b Who plays tennis and hockey but not badminton?

c Who plays tennis, hockey and badminton?

d Who plays only hockey?

e List the members of (T ∪ B) ∩ H′

{ _____ }

f List the members of (T ∩ B)∪ H′

{ _____ }

g Shade in the part of the diagram that shows T′ ∩ H′ ∩ B

h Uri joins the group.
There are now 6 people who do not play tennis, 6 who do not play hockey and 6 who do not play badminton.
Put Uri in the correct place in the Venn diagram.

You can use a Venn diagram to show the number of elements in a set instead of listing all the elements.

1 The Venn diagram shows that $n(B \cap A') = 5$
Complete the Venn diagram to show this information.

$n(B) = 11$

$n(A) = 9$

$n(\xi) = 21$

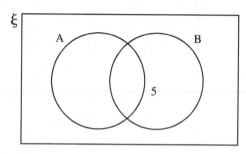

2 Complete this Venn diagram to show:

$n(P) = 12$

$n(Q) = 15$

$n(P \cap Q) = 4$

$n(P') = 16$

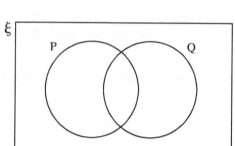

3 There are 30 students in a class.

16 students are male, and 11 students wear glasses.

There are 10 females who do not wear glasses.

a Show this information in the Venn diagram.

b How many females wear glasses?

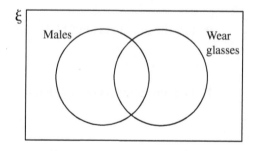

4 In a club with 65 members, 35 take part in archery (A),
29 go bowling (B) and 24 enjoy canoeing (C).

7 members take part in all 3 activities.

3 members do not take part in any activites.

5 members take part in archery and bowling but do not
go canoeing.

3 members go bowling and canoeing but do not take part
in archery.

Use this information to complete the Venn diagram.

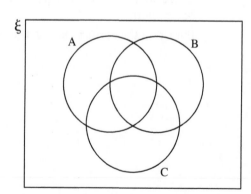

Multiplying a row matrix by a column matrix 🎗

Student book topic 22.1

$$(a \quad b \quad c)\begin{pmatrix} d \\ e \\ f \end{pmatrix} = (ad + be + cf)$$

1 Complete this calculation:

$$(5 \quad 6)\begin{pmatrix} 3 \\ 2 \end{pmatrix} = (5 \times \underline{\hspace{1cm}} + \underline{\hspace{1cm}} \times 2) = (\underline{\hspace{1cm}})$$

2 Find the value of $(4 \quad 2 \quad 1)\begin{pmatrix} 1 \\ -3 \\ 2 \end{pmatrix}$

_____

3 Find the value of $(3 \quad 1 \quad -2 \quad -3)\begin{pmatrix} 1 \\ 3 \\ 2 \\ -2 \end{pmatrix}$

_____

4 $(3 \quad x \quad 2)\begin{pmatrix} 2 \\ 3 \\ 1 \end{pmatrix} = (20)$

Complete this equation and solve it to find the value of x.

$$6 + \underline{\hspace{1cm}} x + \underline{\hspace{1cm}} = 20$$

$$\underline{\hspace{1cm}} x = \underline{\hspace{1cm}}$$

$$x = \underline{\hspace{1cm}}$$

5 Find the value of y.

$$(5 \quad y)\begin{pmatrix} 2 \\ -1 \end{pmatrix} = (7)$$

$y = \underline{\hspace{2cm}}$

6 $(-1 \quad x \quad 4)\begin{pmatrix} 3 \\ 3 \\ x \end{pmatrix} = 11$

Find the value of x.

$x = \underline{\hspace{2cm}}$

7 $(p \quad 2 \quad 0 \quad -2p)\begin{pmatrix} 3 \\ 3 \\ -7 \\ 1 \end{pmatrix} = (15)$

Find the value of p.

$p = \underline{\hspace{4cm}}$

8 Draw a ring around the odd one out in each row.

a $(3 \quad 2 \quad -1)\begin{pmatrix} 3 \\ 1 \\ 5 \end{pmatrix}$ $\qquad$ $(1 \quad 2 \quad -1 \quad 3)\begin{pmatrix} 2 \\ 1 \\ 1 \\ 1 \end{pmatrix}$ $\qquad$ $(5 \quad 1)\begin{pmatrix} 1 \\ -1 \end{pmatrix}$

b $(2 \quad -1)\begin{pmatrix} 3 \\ -2 \end{pmatrix}$ $\qquad$ $(1 \quad 2 \quad -1)\begin{pmatrix} 3 \\ 3 \\ -1 \end{pmatrix}$ $\qquad$ $(4 \quad 2 \quad -3 \quad -1)\begin{pmatrix} 1 \\ 3 \\ 1 \\ -1 \end{pmatrix}$

c $(5 \quad 2 \quad -2)\begin{pmatrix} -2 \\ 3 \\ -1 \end{pmatrix}$ $\qquad$ $(3 \quad 2)\begin{pmatrix} -2 \\ 4 \end{pmatrix}$ $\qquad$ $(-3 \quad -2 \quad -1 \quad 2)\begin{pmatrix} -2 \\ 3 \\ 2 \\ 2 \end{pmatrix}$

9 Explain why you cannot find x when:

$(2x \quad 3 \quad -x)\begin{pmatrix} -1 \\ 2 \\ -2 \end{pmatrix} = (6)$

Multiplying matrices ✿

Each row of the first matrix is multiplied by each column of the second matrix.

1 Complete this calculation.

$$\begin{pmatrix} -2 & 1 \\ 3 & 0 \\ -1 & 2 \end{pmatrix} \begin{pmatrix} 2 & 1 \\ 3 & -1 \end{pmatrix}$$

$$= \begin{pmatrix} -2 \times 2 + 1 \times 3 & -2 \times 1 + 1 \times \underline{} \\ 3 \times \underline{} + \underline{} \times \underline{} & \underline{} \times \underline{} + \underline{} \times -1 \\ \underline{} \times \underline{} + \underline{} \times \underline{} & \underline{} \times \underline{} + \underline{} \times \underline{} \end{pmatrix} = \begin{pmatrix} \underline{} & \underline{} \\ \underline{} & \underline{} \\ \underline{} & \underline{} \end{pmatrix}$$

2 a A is the matrix $\begin{pmatrix} 1 & 2 \\ -1 & 0 \end{pmatrix}$, B is $\begin{pmatrix} 1 & 3 & 0 \\ 2 & -1 & 0 \end{pmatrix}$, C is $\begin{pmatrix} 2 \\ 1 \\ -3 \end{pmatrix}$ and D is $(1 \quad 4)$.

Put a ✓ or a ✗ by these to say whether it is possible to calculate:

i AB _____   **ii** AD _____   **iii** BC _____

iv BA _____   **v** DA _____   **vi** CD _____

b It is possible to multiply DBC.
What would be the order of the matrix DBC?

_____ × _____

3 Calculate

a $\begin{pmatrix} 2 & 3 & -1 \\ -2 & 3 & 0 \end{pmatrix} \begin{pmatrix} 1 & 2 \\ 0 & -1 \\ -2 & 3 \end{pmatrix}$

$\begin{pmatrix} \end{pmatrix}$

b $\begin{pmatrix} 1 & 2 \\ 0 & -1 \\ -2 & 3 \end{pmatrix} \begin{pmatrix} 2 & 3 & -1 \\ -2 & 3 & 0 \end{pmatrix}$

$\begin{pmatrix} \end{pmatrix}$

c $\begin{pmatrix} 2 & 2 \\ -1 & -1 \\ -2 & 2 \end{pmatrix} \begin{pmatrix} 2 & -1 \\ -2 & 1 \end{pmatrix}$

$\begin{pmatrix} \end{pmatrix}$

4 Find the values of the letters in this calculation.

$$\begin{pmatrix} 2 & a \\ b & -1 \\ c & 3 \end{pmatrix} \begin{pmatrix} 4 & 3 \\ 1 & d \end{pmatrix} = \begin{pmatrix} 10 & 4 \\ 3 & e \\ -5 & f \end{pmatrix}$$

$a =$ _____ , $b =$ _____ , $c =$ _____ , $d =$ _____ , $e =$ _____ , $f =$ _____

5 If $(a \quad 3 \quad -1) \begin{pmatrix} 2 \\ 0 \\ b \end{pmatrix} = (0)$, what can you say about a and b?

2 × 2 matrices 🎗

To add matrices, add the numbers in corresponding positions.

To multiply, multiply a row by a column.

1 $A = \begin{pmatrix} 2 & 1 \\ -2 & 3 \end{pmatrix}$, $B = \begin{pmatrix} 1 & -1 \\ 1 & -2 \end{pmatrix}$ and $C = \begin{pmatrix} -2 & -1 \\ 2 & -3 \end{pmatrix}$

Find

a $A + B$

b $B - C$

c AB

d BA

e C^2

f A^2

g AC

h Use your answers to say whether this statement is true or false:

$A + B = B - C$ _____

$AB = BA$ _____

$A^2 = C^2$ _____

$A^2 + AC = \begin{pmatrix} 0 & 0 \\ 0 & 0 \end{pmatrix}$ _____

2 $A = \begin{pmatrix} 2 & 5 \\ 1 & 3 \end{pmatrix}$, $B = \begin{pmatrix} 3 & -5 \\ -1 & 2 \end{pmatrix}$

    a $C = A^2$

      Work out C.

    b Work out CB.

    c What do you notice about your answer to **b**?

    _____

 3 $\begin{pmatrix} 4 & 2 \\ a & -b \end{pmatrix} \begin{pmatrix} b & a \\ -2 & b \end{pmatrix} = \begin{pmatrix} 0 & c \\ 5 & d \end{pmatrix}$

    Find the values of a, b, c and d.

    $a =$ _____ , $b =$ _____ , $c =$ _____ , $d =$ _____